Fossil Collector's Guide to the North Sulphur River

Presented by the *Dallas Paleontological Society*

Authors: *Mark G. McKinzie*
Ron Morin
Ed Swiatovy

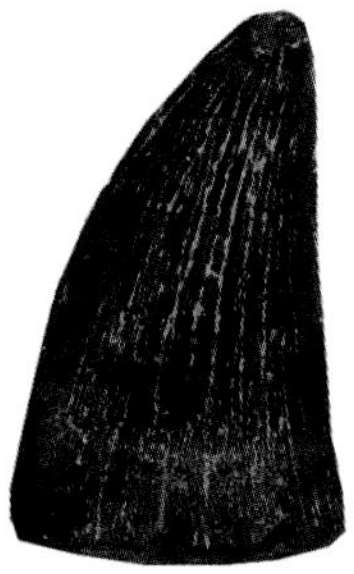

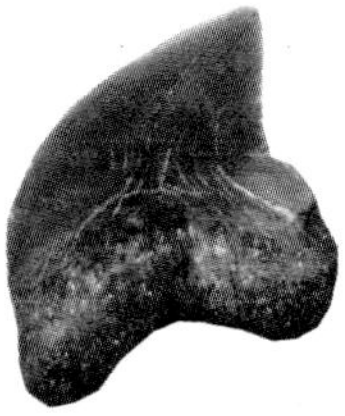

Fossil Collector's Guidebook to the North Sulphur River

Library of Congress Catalog Number:

ISBN 0-9714271-0-0

Printed and bound in the United States of America

About the Authors

Mark G. McKinzie is currently a CAD technician for MCIWorldcom. Previously he was a Geoscience Technician with ExxonMobil Oil Corporation for 12 years in their Oklahoma City and Dallas offices. He is a active member of the Dallas Paleontological Society. A life long lover of fossils in general and Ordovician echinoderms in particular.

Ron Morin is a consulting biostratigrapher who specializes in calcareous nannofossils (2.5 to 25 microns) to determine the age and the characterization of hydrocarbon reservoirs. Prior to becoming a consultant, he was a paleontological advisor for Mobil Oil Corporation where he worked for 30 years in their Los Angeles, Dallas, and New Orleans offices. He began collecting megafossils in the Dallas-Fort Worth area in the early 1970's and focused on ammonites because of their age-dating potential.

Ed Swiatovy is a retired Engineering Consultant to the telecommunications wire and cable industry. He is an active member of the Dallas Paleontological Society and, in concert with his wife Catherine, an avid collector of Cretaceous marine fossils.

Abstract

The North Sulphur River of northeast Texas has long been known to be a fossil collector's treasure trove for Late Cretaceous and Pleistocene vertebrate and invertebrate fossils alike. In its upper drainage, the original river channel was the north-south boundary between Hunt and Fannin, and Delta and Lamar counties. For flood control and agricultural purposes, it was dredged and straightened, exposing bedrock of the Ozan, Wolfe City, and Pecan Gap Formations. Erosion in the years since has produced deep cuts within the river channel itself and the tributaries feeding into it.

This report is a summary of the geology and paleontology of the North Sulphur River. The author's efforts are focused along the first twenty linear miles of the river channel. This area extends from the North Sulphur River's headwaters to just east of the Highway 24 bridge on the Paris to Commerce road.

The Dallas Paleontological Society hosted a fossil Symposium at the Dallas Museum of Natural History at Fair Park in Dallas, Texas on Saturday, February 13th, 1999. The Society invited the general public and all interested parties to attend. The purpose of the Symposium was to capture on basemaps and photographs, the rich fossil fauna and flora collected by the amateur fossil community of Texas and adjacent states. The ultimate goal of the Dallas Paleontological Society was to publish a guidebook to the fossils of the North Sulphur River, including an extensive photo identification section.

Acknowledgements

First of all, on behalf of the entire Dallas Paleontological Society (DPS), let me express our sincere thanks to all the amateur fossil collectors who took the time and effort to attend the North Sulphur River (NSR) Fossil Symposium on February 13th, 1999. Over 130 fellow fossil collectors paticipated in the Symposium. It is *your* collecting efforts over the years that has brought to light most of the significant fossil finds from the North Sulphur River (NSR). And it is your fossil specimens that make up the bulk of the photo identification material in this guidebook. This book really is done for, and dedicated to, the amateur fossil community of Texas and beyond.

Second of all, the DPS wishes to express their gratitude to the staff of the Dallas Museum of Natural History for opening their doors (literally!) to the Society for the purpose of hosting the Symposium. Special thanks go to Tony Fiorillo, Chief Paleontologist at the museum, Robert Townshend, Museum Director and Nora. The NSR Symposium could not have been possible without the use of this facility.

Finally, and most importantly, I would like to thank all the DPS members who volunteered to make the NSR Symposium and this guidebook a reality. Without your help, none of this would have been accomplished. I especially want to salute the following people. Chris Vencevich photographed the fossils and scanned specimens during the symposium. Ron Morin, an ammonite expert, helped with the time correlation and ammonite zonation of the rocks involved. Ron also wrote Chapter 3 on Cretaceous cephalopods. Edward Swiatovy wrote Chapter 5 on the Cretaceous sharks and rays of the North Sulphur River. Richard Zack spent the past four years quietly compiling an exhaustive bibliography of publications concerning NSR geology and paleontology. Ingrid Center helped in the initial planning stages. Bill Lowe, Roger Farish and Cathy Swiatovy did the tedious task of proofreading. And, of course, a special thanks to all the volunteers at the North Sulphur River Symposium who helped staff the tables, identify the fossils, and tabulate fossil location data. These included Richard Benefield, Brita Bishop, Brian Bowles, Jim Chaney, Roger and Linda Farish, Rocky Manning, John and Anne Meyers, Ron Morin, Ed and Cathy Swiatovy, and Velma Urbano-Garza.

A special thanks to Keith Young, who was cornered into identifying ammonites for the first hour of the symposium. Dr. Gale Bishop at the University of Southern Georgia helped with the identification of the Cretaceous decapods. Dr. Roy Enrico with ExxonMobil Oil Corporation identified the forams from the basal Ozan Formation. I would also like to thank Brita Bishop (1998 DPS president) and Anne Meyer (1999 and 2000 DPS president) for allowing me the opportunity to pursue and complete the North Sulphur River project.

This project is in part supported by the Barry Leon memorial fund. Deceased DPS member Barry Leon discovered a partial adult mammoth on the NSR while on a Society field trip. This was a significant contribution to northeast Texas paleontology and resulted in a month long cooperative excavation between the DPS and the Dallas Museum of Natural History.

In all the photographic plates, the names of the collectors who discovered the fossils are listed for each Figure. In Chapter 5, all photographed specimens are from Ed Swiatovy's collection unless otherwise stated..

The draft was prepared for publication using Pagemaker 6.5 by consulting geologist John McLeod. In addition, Jan Collmer donated $2500 to offset the printing costs of this book. Thank you Jan.

Table of Contents

Introduction

The North Sulphur River of northeast Texas has long been known to be a fossil collector's treasure trove for Late Cretaceous and Pleistocene vertebrate and invertebrate fossils alike. Exceptional specimens of nearly complete mosasaurs and several species of fish have been recovered. In addition, an abundant variety of shark's teeth, ammonites, and other molluscs can be collected from its gravel bars and riverbed. Indeed, after a heavy rain, it is not uncommon for an army of dedicated fossil collectors to be found tromping through the receeding waters, searching for that special find. Nothing is more disappointing than discovering footprints ahead of you on a gravel bar when you have gotten there at the crack of dawn. Who are these people anyway? The readers of this book, that's who!

The North Sulphur River is a generally west-to-east flowing stream with its headwaters northeast of Bailey in Fannin County. The South Sulphur River links up with it at the junction of Lamar, Delta, Hopkins, Franklin, and Red River Counties. The combined Sulphur River continues eastward where it eventually feeds into the Red River south of Texarkana in southwest Arkansas. In the NSR upper drainage, the original river channel was the north-south boundry between Hunt and Fannin, and Delta and Lamar counties. For flood control and agricultural purposes, it was dredged and straightened in 1929, exposing bedrock of the Ozan and Wolfe City Formations in its upper portions. Erosion in the years since has produced deep cuts within the river channel itself and the tributaries feeding into it. If it had not been for this man made channelization, there would not be the extensive exposures of fossil-bearing rocks currently seen today.

This report is a summary of the geology and paleontology of the upper portions of the North Sulphur River. Our efforts are concentrated along the first twenty linear miles of the river channel. This area extends from the North Sulphur River's headwaters to just east of the Highway 24 bridge on the Paris to Commerce road. There has never been a guidebook dedicated to the fossils of the North Sulphur River available to the general public. A tremendous amount of work has been published on various fossil groups (both microfosssil and macrofossil) in assorted professional publications over the years. Not to mention the number of unpublished master's theses and doctor's dissertations from universities in Texas alone. The Dallas Paleontological Society has been compiling an exhaustive bibliography of articles related to the North Sulphur River for years. This guidebook is an attempt to capture all that hard to get to data and summarize it in a easy-to-read format for the casual reader. Emphasis will be placed on the fossils as the fossils are the reason for the guidebook in the first place.

In connection with this, the Dallas Paleontological Society hosted a fossil Symposium at the Dallas Museum of Natural History at Fair Park in Dallas, Texas on Saturday, February 13th, 1999. The Society invited the general public and all interested parties to attend. The purpose of the Symposium was to capture on basemaps and photographs, the rich fossil fauna and flora gathered by the amateur fossil-collecting community of Texas and adjacent states. The Symposium was a success beyond this humble collector's wildest imagination. Over 130 participants brought in a incredible array of Cretaceous and Pleistocene fossil material. Many of the specimens are new to science, and will have to wait for formal scientific description. As a result of the Symposium, the Dallas Paleontological Society can offer a comprehensive photo identification guide to the fossils of the North Sulphur River.

Is this the final word on the fossils of the North Sulphur River? Absolutely not! Believe it or not, paleontology is a dynamic science. New discoveries constantly answer old questions, and ask new ones. The fossil *you* find tomorrow may solve a riddle of yesterday. Consider this a snapshot of what is currently known in the professional and amateur communities alike. The only thing that is guaranteed is that some of the information presented herein is out of date by the time the book went to press. Be that as it may, this is and will be a valuable reference source on the fossils of the North Sulphur River.

A basic understanding of the Late Cretaceous geology of northeast Texas is needed to fully appreciate the fossils collected along the North Sulphur River. The North Sulphur River (NSR) is located in the extreme northeast corner of Texas (see Fig. 1). NSR will be used interchangeably as an abbreviation for the North Sulphur River throughout the remainder of the book. It is relatively flat topography of broad, east-west valleys separated by gently rolling hills. When approaching the NSR from any of the numerous bridge access points, you usually don't see the river channel until you are almost on top of it. The present-day topography is in large part controlled by the underlying structural geology and the drainage patterns established since the last Ice Age (Pleistocene).

The age of the Earth has been subdivided into a geologic time scale based on the observation of rock sequences around the globe over the last 300 years. Modern geology had its roots in England in the 19th century when James Hutton and other geologists recognized the laws of stratigraphy. Basically, unaltered rock units are layered vertically much like a stack of pancakes with the oldest ones on the bottom, and the ones on top getting progressively younger as you move up the column.

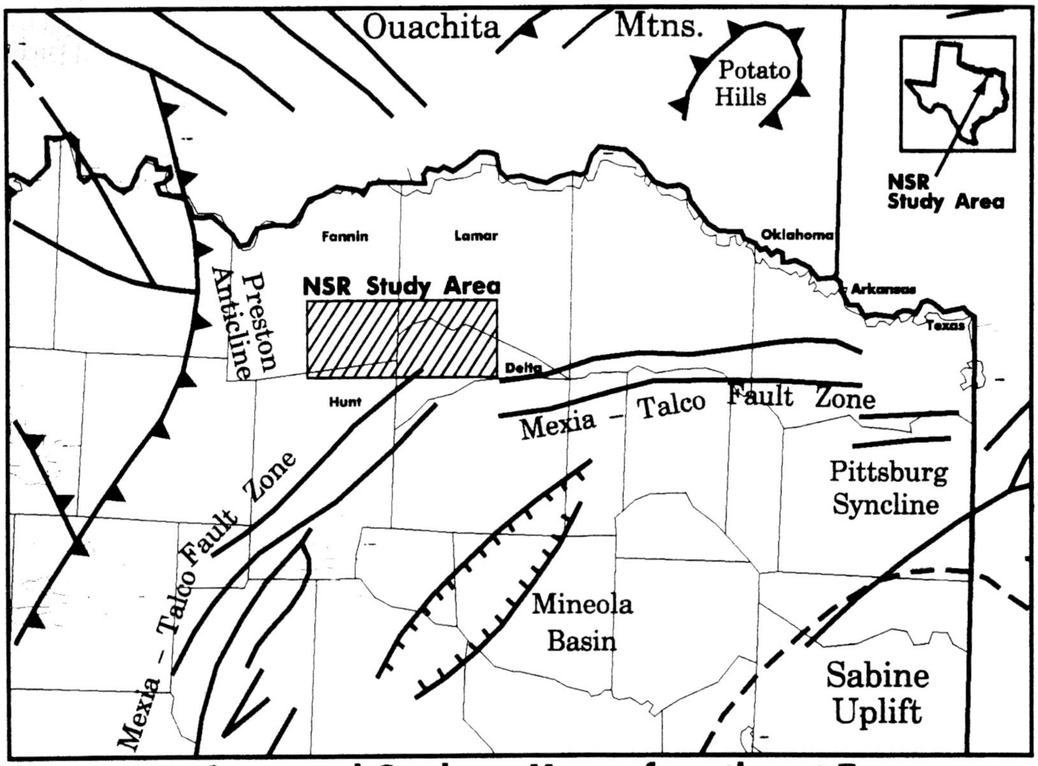

Figure 1. – Structural Geology Map of northeast Texas
Source: Geological Highway Map of Texas (1965) Exxon World-wide Tectonic Map (1994) Stehli et al (1972)

As obvious as it seems to us today, there was a time when no one recognized the fact that in a sequence of horizontal rock beds, the oldest ones would naturally be on the bottom. Prior to the early 1800's, all the sediments and rocks on the surface of the Earth were attributed to the Great Flood of the Old Testament in the Bible. Imagine that a cook has placed blueberries in one batch of pancake batter. He then inserts this into the stacks of plain pancakes as he serves them. If you had two plates full of pancakes side by side, you could correlate the two by connecting lines between the layers with the blueberries in them. Those pancakes below the one with the blueberries would be older than the ones above the blueberry level. You have now determined the *"relative"* age of the pancakes. You still don't know what time the cook made the blueberry pancakes (8AM, 2PM, etc.) or the calendar date. Consider fossils to be the blueberries in the pancake mix, and the pancakes are equivalent to the matrix comprising a rock unit.

Yet you still don't know what time the cook made the blueberry pancakes. This is called the ***"absolute"*** age. For that, we need some other kind of ingredient in the batter. Something that is essentially "frozen" in time so that it recorded the date it was added to the batter. Luckily, in nature we find numerous elements that have atomic clocks associated with them , these help to ascertain the absolute age of a formation. The most common involve radiometric-age dating of potassium (K) – argon (Ar) isotope ratios. For very young deposits (geologically speaking) of 1/2 million years or less, carbon based dating is used. Organic based carbon decays after burial from one form (isotope) to another and this rate has been plotted versus time. Bentonites are deposits of volcanic ash that have settled on the sea floor. They are ideal beds for radiometric-age dating techniques due to the microscopic zircon and sanadine crystals within them. These minerals formed within the magma chamber of the volcano prior to its eruption. As the magma cooled and the minerals precipitated out, their atomic clock is set to zero and starts ticking

Another popular method for reconstructing the Earth's history is magnetostratigraphy. Certain mineral in the rock matrix can be measured by their magnetic polarity signature. Oceanic crust is especially applicable to this procedure. Throughout geologic time, the magnetic poles of the Earth have flip-flopped. If you use a compass today, the needle points north. During a magnetic pole reversal the compass needle would point south. Certain minerals behave like the compass needle when they are formed and permantely "point" in the direction of magnetic north at that time. Geoscientists have recorded in detail the magnetic pole reversals extending back to the Late Jurassic. These measurements have been correlated to the geomagnetic polarity time scale (GPTS).

Unfortunately, not every rock unit has any distinguishing markers like this. Then geologists rely on the fossil record to relatively age date one formation to another. It gets even more complicated when there are no fossils or marker elements to date a formation. All the above mentioned dating procedures form the basis for the geologic time scale used globally today.

The formations exposed along the NSR study area date from the Mesozoic Era which is Latin for *middle life*. It followed the Paleozoic era (*early life*), and preceded the Cenozoic (*modern life*). It was the age of the dinosaurs. The Mesozoic is further subdivided into the Triassic, Jurassic, and Cretaceous periods in ascending order. Each period is further subdived into lower(early), middle, and upper(late) epochs. Each epoch is further refined to European stages based on European outcrops where *index fossils* have been collected and a "type section" established. Ideally, an *index fossil* is a fossil with wide geographic range (hopefully worldwide!), and a narrow stratigraphic range (1-2 million years at most). This makes an index fossil an ideal blueberry in our pancake stack for relative-age dating. Most of the Late Cretaceous type sections are from the thick chalk sequences along the coast of France where a majority of the ammonite zonation was originally done.

During the Cretaceous, especially in the Late Cretaceous, the marine ammonites underwent rapid evolution. Many chronologic species of short-lived duration (under 1 million years!) evolved, lived, and went extinct in basins all over the world. This makes them excellent candidates for index fossils. That is the good news. The downside is that many ammonites were endemic to a certain sea or region of the world at that time. That means they would have been restricted to say the Indian Ocean or the Gulf of Mexico or the Arctic Ocean of today. So, the ammonites were very good <u>regional</u> index fossils but not necessarily good <u>global</u> index fossils.

Recognizing this limitation, ammonites are still the index macrofossil of choice for correlating Late Cretaceous marine deposits around the world today. They are used in conjunction with other fossil groups to correlate formations from one part of the United States to another or from one continent to another.

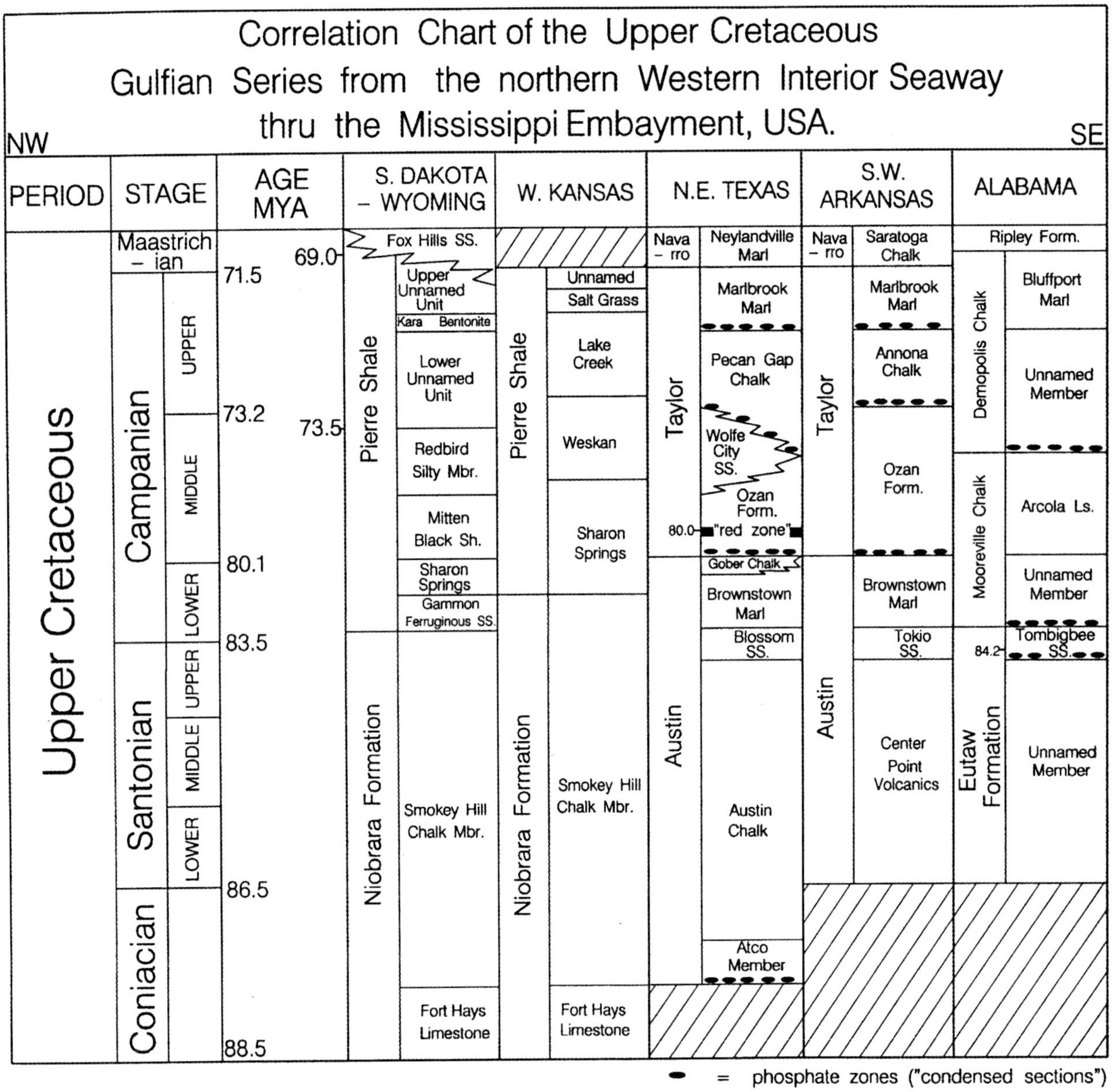

Figure 2. – Regional Upper Cretaceous Correlation Chart

Data for this correlation are based on the following: Young et al., 1976; McNeil and Caldwell, 1981; Gill and Cobban, 1966; Kauffman, 1977; Gill et al., 1972; Smith and Mancini, 1983; Bottjer, 1986; Kennedy and Cobban, 1993; Cobban and Kennedy, 1993; Mancini and Soen, 1994; Kennedy et al, 1997.

Fig. 2 is a regional correlation chart of the Late Cretaceous deposits across the United States. It is based upon the work of many different geologists trying to correlate different formations from different parts of the country to one another. During the time interval we are concerned with, the present-day Gulf of Mexico was the southern end of a vast inland sea that extended all the way from Alaska and Greenland. This ocean is called the Western Interior Sea (see Fig. 6). It existed throughout most of the Cretaceous with minor transgressions and regressions primarily related to the tectonic activity of the emerging ancestral Rocky Mountains. The shoreline and the maximum depth of the sea fluctuated in response to the mountain building events occurring along the western margin. These included episodes of volcanic activity which periodically spewed ash clouds. These ashfalls drifted eastward across the seaway and settled on the seafloor to eventually become bentonites. These bentonite beds can be traced across hundreds of square miles in the northern portion of the seaway.

One of the primary fossil groups used to tie these sections together is the ammonites. The Pierre Shale sequence of eastern Montana, eastern Wyoming, South Dakota and extending southward into western Kansas is locally pregnant with ammonites. In addition, it has numerous bentonite layers imbedded within it. This makes this area an ideal interval for both relative-age dating (ammonites) and absolute-age dating (bentonites). It has been the focus of intense study for years and the evolution of the ammonites is well established with precise absolute age bracketing due to the bentonites. In fact, the ammonites of the Western Interior Basin are the standard zonation that other parts of the United States are compared to. For an excellent summary of the current state of knowledge of the Pierre Shale (and lots of good ammonite pictures!) see Larson et al 1997 "*Ammonites and the other Cephalopods of the Pierre Seaway*".

Please keep in mind that the regional correlations shown in Fig. 2 are tentative at best and subject to change. As new specimens are collected in the various basins or outcrop areas, minor revisions may occur. Formation X may turn out to be Late Santonian and not Early Campanian based on new fossil discoveries or a better absolute-age dating of some sample. This in turn causes the correlation with the nearest basins to be scrutinized. That is why it is so important for amateur fossil collectors to keep accurate and detailed locality information with each specimen. Only then does it have scientific value as far as the academic community is concerned. There is still not total agreement on the ages of some of the Taylor Group formations in Texas. Ammonites and other macrofossils give one age assignment, while the microfossils indicate another. There are still issues to be resolved here in our own backyard!

All the Late Cretaceous formations from the Coniacian stage and younger in northeast Texas are marine deposits. They were all formed on the sea floor; some closer to the ancient shoreline than others and while some formed in deeper water. That is why we discover marine mosasaur remains and not terrestrial duck-billed hadrosaurs in the same strata. The chances of a land-based dinosaur carcass floating out to sea, settling to the bottom and being preserved were slim indeed. There were too many mosasaurs, scavenging sharks, and predatory fish around to insure that a large carcass would be dismembered and consumed before it ever reached the sea floor. And that is before the microorganisms went to work. Nature is very efficient at recycling organic matter under most conditions.

One of the things to note on Fig. 2 is that the Coniacian and Santonian stages are dominated by chalk sequences. In the northern portion of the Western Interior Seaway it is the Niobrara Formation. In Texas it is the Austin Chalk which is famous to all local collectors for the abundant shark's teeth in the "condensed section" of the basal Atco Member. Late Cretaceous condensed sections are characterized by sand-sized glauconite grains, phosphatised nodules of varying sizes from sand to pebble, remanie fossil molds, and abundant oyster valves. They are also usually rich in vertebrate remains, especially shark's teeth. Any, or all, of these elements may be present in a condensed section.

How are condensed sections formed? Petrographic studies on modern sea floor sediments indicate that the mineral glauconite forms only in areas of slow marine sedimentation. Recent phosphorite deposition occurs in "reducing" marine environments that are rich in organic matter. A reducing environment indicates that there is not enough oxygen in the bottom seawater or at the sediment/water interface to support life. That means that the bacteria would normally process (or consume) any organic matter that drifted to the seafloor cannot survive. Neither can the infaunal decapod crustaceans and polycheate worms that normally churn up the sediment (and the organic matter) during their burrowing activities. That is good news if you want to become a fossil because your odds of being buried and preserved are better if there are no organisms around to process your remains.

But wait a minute! How do you explain the macrofossils that you do find in a condensed section? There are multiple lines of evidence to indicate that the fossils (vertebrate remains, phosphatized mollusc shells, and oyster valves) were transported in from other environments. In other words, the fossils we collect in the condensed sections today died somewhere else, settled to the seafloor, and were later scooped up before burial to the present location. The most likely mechanism for this would be intermittent scouring by storm surge waves. Hurricanes or severe offshore storms would create powerful storm surge waves that would scour the seafloor, picking up living and dead organic material. As the surge moved further offshore, it would decrease in velocity to the point that its sediment load would drop out to the seafloor. Along parts of the seafloor, the surges can deposit debris into areas that would normally not recieve sedimentation. There the mollusc shells would become phosphatized. When the next storm surge wave hit it would deposit new material and rework and abrade shells and vertebrate material already on the seafloor. Multiple generations of this process over time would slowly build up the condensed sections we find today.

Chalks themselves are indicators of low-energy, clear water bottom conditions on a shelf environment. They are composed of biogenic carbonate in the form of billions and billions (to steal a quote from Carl Sagan) of calcareous tests (shells) of nannofossils, and benthic and planktonic foraminifera settling to the seafloor upon death. Imagine this transpiring for countless generations over the years. Eventually you build up enough volume to produce chalk with progressive burial and lithification. Chalks of the Austin Group in Texas consist of pure chalks and rythmically bedded chalk/marl sequences. Marl is an impure chalk that contains a significant amount of mud. The Austin Chalk represents the maximum marine transgression during the Late Cretaceous in northeast Texas.

The Cretaceous was a time of maximum sea level transgression worldwide. In fact, except for the Ordovician period, the world's oceans indunated more of the continental landmasses during the Cretaceous than at any other time since the Precambrian. This was in part due to a major "greenhouse" effect occurring globally at the time. There is little evidence of polar icecaps at either the South or North poles. Water that normally would have been frozen in the polar ice sheets was now part of the world's oceans. We will go into this in more detail later when we discuss Late Cretaceous paleogeography.

As stated previously, increase the shale content of chalk and you produce marl. Shale consists of very small clay-sized particles that drifted out to sea on the ocean currents. It could have come from very distant volcanic eruptions as ash or more likely as sediment carried by river systems into the sea. The Mississippi River today is transporting tons of sand, silt, and clay-sized particles into the Gulf of Mexico every year. In fact, the birdsfoot delta at the southern end of the Mississippi River in Louisiana is a prograding complex of just such sediments that is actively moving out to sea.

The younger Campanian stage formations in Fig. 2 have an increased siliciclastic content compared to the underling Austin Chalk. These are the formations we are dealing with along the NSR study area. Siliciclastics are the clay, silt, and sand-sized particles that are products of erosion of existing rock formations. They are generally of terrestrial orgin but can end up in a marine environment due to fluvial transportation out to sea.

Fig. 3 is the local stratigraphic column/time correlation chart for the NSR study area. The graphic also includes the relative sea level curve versus present day. What is a sea level curve and how is it calculated? If present-day sea level equals zero, then sea level throughout that time interval of the Late Cretaceous was 175 - 250 meters higher than today. There are just over three feet to one meter. That means sea level was up to 820 feet higher than today (all other things being equal)! It also means large portions of Texas were underwater during the Late Cretaceous. If you lived in Grapevine Texas during the slightly older Woodbine deposition you would have had seashore property, not lakeshore. Of course, you would have also had those pesky dinosaurs to deal with as evidenced by the dinosaur tracks on the north shore of Lake Grapevine.

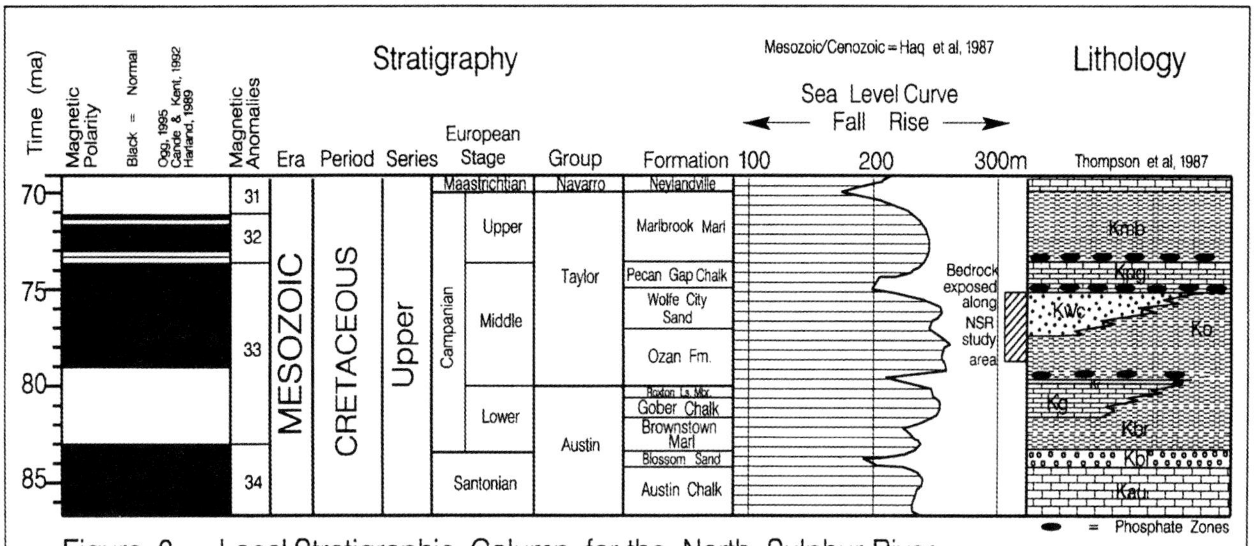

Figure 3. – Local Stratigraphic Column for the North Sulphur River

In northeast Texas the Austin Group, is composed of the basal Atco Member, the Austin Chalk, the Blossom Sandstone, the Brownstown Marl, the Gober Chalk, and the Roxton Limestone Member of the Gober Chalk. This is overlain in turn by the Taylor Group that consists of the Ozan Formation, the Wolfe City Sandstone, the Pecan Gap Chalk, and the Marlbrook Marl. On top of this is the Navarro Group of Maastrichtian age. Naturally, the layers on top are the youngest.

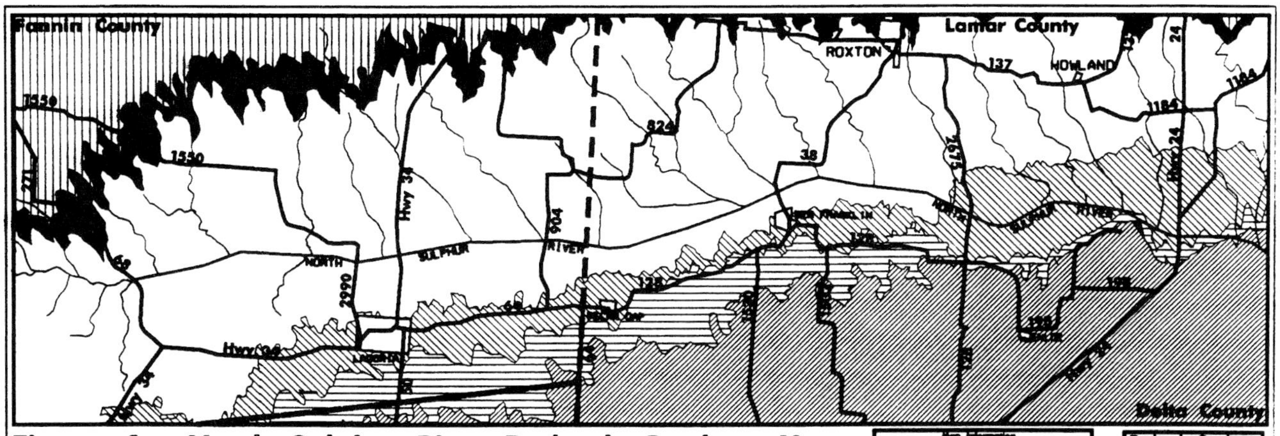

Figure 4. – North Sulphur River Bedrock Geology Map

Fig. 4 is a bedrock geology map of the NSR study area. A bedrock geology map has all the recent surficial deposits stripped away such as vegetation, soils, Quaternary terrace deposits, and any fluvial sediments such as modern day gravel bars. All that is left is the underlying "bedrock" geology. As can be seen in Fig. 4, the NSR in this study area lies within the Ozan, Wolfe City, and Pecan Gap formations which are Middle Campanian in age. I follow here the terminology used by Barnes in the 1966 Geologic Atlas of Texas - Texarkana sheet which mapped the upper unit of the Brownstown Marl as the Ozan Formation. This is also the terminology followed by Cobban and Kennedy who have done much of the published ammonite zonation in the NSR study area.

The exact age of some of the vertebrate material coming from the NSR has been a concern of geoscientists for years. This pertains to the loose vertebrate material found in the gravel bars along the length of the NSR versus material found in situ. Are they all Campanian in age or is it possible that some of it is reworked older Santonian material? Is there an easy way to distinguish between the older, reworked debris and freshly eroded material?

Fig. 5 is an expanded areal coverage bedrock geology map with an overlay of the outline of the NSR drainage basin on it. Anything within the heavy black outline has drainage feeding into the NSR. Anything outside of the heavy black outline drains into the river systems to the north or south. The map clearly indicates that the NSR in its upper drainage erodes only rocks of Early to Middle Campanian age. Only in the extreme northeast corner of the map are creek tributaries reaching into the Late Santonian Blossom Sandstone. Collectors can be fairly certain that anything recovered west of the Highway 24 bridge north of Cooper is Campanian in age.

There is still the possibility that a weathered and tumbled shark's tooth could be older, say Santonian in age. This would imply that it was deposited in the Late Santonian, eroded out in a later scouring event thousands to a few million years later, and redeposited in the basal Ozan Formation condensed section. The tooth was then eroded out again in Recent times by the NSR channel downcutting and redepositing it downstream in a gravel bar. These teeth should look tumbled and smooth due to the multiple episodes of excavation, transport, and deposition. As my wife would say; "they are missing some of their bits and pieces." Santonian age fossils will be the exception - not the rule – in the NSR study area.

As indicated in Fig. 3, and can be seen on the bedrock geology map of Fig. 4, the three formations we are most concerned with are the Ozan Formation, the overlying Wolfe City Sandstone and the Pecan Gap Chalk. Let us briefly review the history and description of these three rock units.

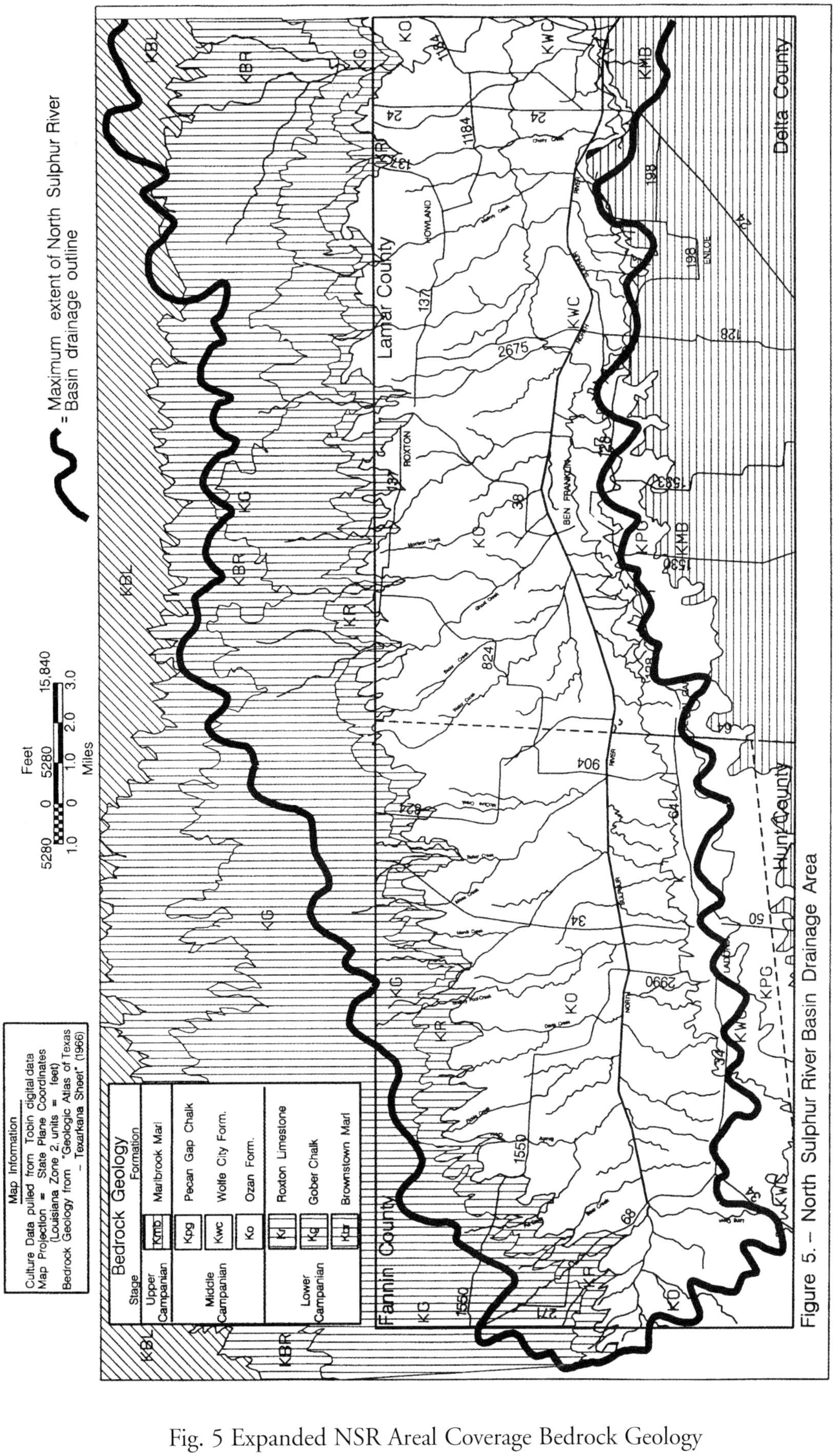

Fig. 5 Expanded NSR Areal Coverage Bedrock Geology

OZAN FORMATION

The Ozan Formation was first cited as part of the overlying unit above the Annona Chalk at White Cliffs Landing on the Little River, Sevier County, southwest Arkansas in 1888 by Hill. The type locality is from outcrops along the middle fork of Ozan Creek in Hempstead County, Arkansas. It was originally considered the upper part of the Brownstown Marl by Veach in 1905. It was formally described by Dane (1929, p. 58) in Bulletin 1 of the Arkansas Geological Survey. In Arkansas, it consists of of a sandy, micaceous (shaley) marl. In its basal part there is a glauconitic sand to sandy marl termed the Buckrange Sand Lentil by Dane (1929). It is up to 200 feet thick, and disconformably overlies the Brownstown Marl. The upper contact with Annona Chalk in Arkansas is conformable and gradational in places.

In Texas, the Ozan Formation consists of up to 425 feet (130 meters) of bluish-gray, calcareous clays and mudstones with occasional thin, sandy layers. It weathers easily into smooth, rounded plates of light-gray color shale along the riverbanks. Some of them have a nodular appearance. Marine megafossils are common and scattered thruout the formation. However, it takes stream erosion and redepostion in the river gravel bars to concentrate the fossils. The basal portion contains a condensed section characterized by abundant phosphate nodules of all shapes and sizes.

Of particular interest to fossil collectors is the informal "red zone" that outcrops west of the Highway 2990 bridge north of Ladonia. The red zone extends from two/tenths of a mile upstream from the 2990 bridge to a point approximately 1.4 miles on both sides of the river. It also occurs in the bed of Davis Creek and a unnamed tributary across from it on the south side of the river. The best exposures are at the mouths of these two creeks (see Plate 1, Fig.s 5-7). This zone has long been known to fossil collectors for its abundant ammonites and other molluscs. The fossils come from a distinctive, thin, glauconitic unit that overlies a red-weathering condensed section. The condensed section is loaded with black phosphate nodules, many of which are moulds of various molluscs. The red zone has an overall nodular appearance with the fossils weathering out of the matrix. This unit occurs 151 feet (46 meters) above the underlying Gober Chalk.

The Ozan Formation contact with the underlying Gober Chalk has been described as both conformable and disconformable in the literature. The type locality for the Gober Chalk was designated by Stephenson (1927) from outcrops along Allen Creek at the village of Gober in south-central Fannin County. The formation can be observed in scattered exposures from southwestern Fannin County to as far east as Clarksville, Texas where it eventually pinches out into interbedded marl layers of the Brownstown Marl and the overlying Ozan Formation. The Gober Chalk is the uppermost formation of the Austin Group and the upper tongue of the Austin Chalk proper. Lithologically, it is indistinguishable from the Austin Chalk as seen along the "white rock" escarpment in the Dallas area.

In the NSR study area it consists of buff to cream argillaceous chalk which weathers into small white flags. Macrofossils are scarce and scattered throughout. They include large *Inoceramus* shells, very small brachiopods, *Hamulus* worm tubes and occasional echinoid spines. The chalk was deposited under relatively clear water conditions during the peak of a transgressive cycle.

The uppermost portion of the Gober Chalk is the Roxton Limestone Member. This unit is up to 3 meters thick and is an extremely well-indurated bioclastic grainstone to packestone. It is very tough and forms small benches and waterfalls. It has been historically used as a building stone. It can be seen along drainage ditch exposures at the junction of Highways 38 and 137 in the town of Roxton in Lamar County north of the NSR.

Contact exposures between the Gober Chalk and the Ozan Formation are few and far between. At the type locality in Allen Creek in the town of Gober, the chalk to marl lithologic contact is sharp and uniform. There are no sedimentary features indicative of a depositional hiatus. However, at its easternmost extent, the contact is a 5cm thick calcarenite rich in phosphate nodules and organic debris. In the NSR study area, there is a good contact exposure that is of special interest to fossil collectors. It is located north of the village of Bug Tussle (junction of Highways 34 and FM 1550 north of Ladonia) where Highway 34 crosses a tributary feeding the NSR. The contact is exposed on the downstream side of the creek under the bridge.

The bridge foundation is resting on Gober Chalk. A small bench and waterfall has formed where the Ozan Formation has been downfaulted relative to the Gober Chalk. Plate 1, Fig. 3 is a view looking westward at the upthrown Gober Chalk at the base of the bridge. The fault strikes southwest – northeast and the vertical displacement is greater than one meter. Evidence of the fault can be seen in a 1 – 2 inch thick stylolite seam that parallels the fault trace. The geologic term **STYLOLITE** refers to a diagenetically formed solution surface caused by differential movement under pressure. Say what? In mechanical terms, it is a carbonate rock's response to the stresses associated with faulting. In this case, the stylolite cavity has been replaced with prismatic calcite that can be seen in cross-section and collected downstream as weathered-out chunks.

On the down thrown side of the fault, the top of the Gober Chalk is exposed as a hardground surface. It is extensively burrowed and iron oxide stained giving it an overall rust color and hummocky appearance. Both of these features are indicative of a period of non-deposition in which the sea floor semi-lithified and was actively reworked by infaunal organisms. None of the sea creatures that caused the burrowing are preserved as fossils. However, the burrows can probably be attributed to decapods, echinoids, and infaunal pelecypods.

Immediately above this hardground is a thin (less than one foot) variable thickness horizon characterized by its deep red color, black phosphate grains, and abundant micro shark, ray and fish teeth. Plate 1, Fig. 4 is a closeup view of the basal Ozan Formation onlapping the top of the Gober Chalk. Hundreds of complete microteeth can be collected from this site in a five gallon bulk sample! A majority of the teeth are on the order of 3mm or less in size. Other vertebrate remains include shark and bony fish corprolites and vertebrae. There are also small reworked calcareous fossils from the underlying chalk in this deposit. These include foraminiferans, *Hamulus* worm tubes, terbratulid brachiopods, crinoid root and stem fragments, broken echinoid spines and bryozoan and coral fragments. The only large shark remains are broken blades of larger *Squalicorax* teeth.

Capping this unit is a red siderite nodule zone originating from the iron-rich, sediment filled burrows of infaunal organisms. This upper surface looks much like the decapod-rich nodule zones from the Lake Lewisville Spillway in Denton County. Those occur in the Britton Formation of the Eagle Ford Group and probably formed under similar environmental conditions. One beautifully preserved heart urchin was collected *in situ* from the top of the red nodule layer (see Plate 4, Fig.s 1-3). The amount of microteeth and phosphate pebbles decreases upwards from the base of this unit and its contact with the underlying Gober Chalk hardground. Overlying the red layer is the classic gray marl of the Ozan Formation.

The basal unit of the Ozan Formation at this locality is somewhat of an enigma. It clearly represents the initial regressive deposit of the Taylor Group sequence on top of the Austin Group Transgression. Transgressions represent periods of marine onlap on continental shelves and are dominated by carbonate sequences (limestones, dolomites, and chalks). Regressions represent periods of continental onlap over the shelves and an increase in sand, silt and shale content of the marine formations. These result in sandstone, shale and marl deposits.

This microtooth zone is probably of limited aerial extent. It may be the result of a local depression where teeth were concentrated into a slight depression by storm surge currents during a single depositional event. The small size and excellent preservation of the teeth (very few of the microteeth show signs of abrasion) indicate that they could not have been transported far and were quickly buried. The larger *Squalicorax* teeth are probably reworked teeth that have traveled greater distances. This local depression may have been only a few feet lower than the surrounding seafloor and we are observing the very distal deposit of a storm surge event.

Another possibility is that this site represents a shark breeding and/or nursing area. Sharks tend to congregate by size for protection from other predators (including other sharks). At the time of deposition, the immediate vicinity may have offered ideal conditions (shelter, food source, etc.) for juvenile or immature sharks before they were big enough to venture into deeper open waters. The sedimentological evidence is not conclusive for or against this hypothesis.

The Ozan contact with the overlying Wolfe City Sandstone is conformable. In the few places where a contact is observed there is usually a series of transitional beds. Along the NSR it is not well developed and is easily overlooked. The contact lies upstream from the FM 2675 bridge east of Ben Franklin.

WOLFE CITY SANDSTONE

The Wolfe City Sandstone was originally described by Stephenson (1918, p.155) who designated the type locality for a railroad cut just northeast of the town of Wolfe City in Hunt County, Texas. Unfortunately, only the middle section of the formation is exposed there. The formation consists of 23-25 meters of calcareous gray sand to sandy marl and mudstones. At the type locality, there is a thin (30-75 cm thick), persistent, concretionary sandstone layer. It is predominately composed of molluscan fossil molds, especially oysters. Ichnofossils are common and represent the burrowing and foraging activities of an active and diverse infaunal assemblage. The Wolfe City Sandstone is the only sand facies within the Taylor group. It consists of two distinct lithofacies per Richardson (1972, p. v) which Thompson et al (1991, p.19) have since been demonstrated to be lateral equivalents. The sandy units are friable to well-cemented, very fine-grained and exhibit cross-laminations to faint horizontal laminations. The sand grains are immature, glauconitic, chert-bearing and subarkosic. Grain size increases northward. The Wolfe City Sandstone gets progressively thinner as you move east from the type section towards Paris. In the NSR study area, the Wolfe City Sandstone is overlain by the Pecan Gap Chalk. The contact is disconformable and marked by a thin condensed section of abundant phosphate nodules, remanie fossils, and vertebrate remains.

PECAN GAP CHALK

The final Cretaceous unit is the Pecan Gap Chalk. The NSR cuts across it approximately a half mile downstream from the Highway 24 bridge in Delta and Lamar Counties. The Pecan Gap chalk was originally described by Stephenson (1918, p. 156) from exposures along the Gulf, Colorado & Santa Fe Railroad a half mile east from the town of Pecan Gap in Delta County. Only the lower ten feet and the unconformable contact with the underlying Wolfe City Sandstone is exposed here. A good exposure of the Pecan Gap Chalk on the NSR occurs along the Slabtown – Pacio low water crossing road 3 miles downstream from the Highway 24 bridge.

In Hunt, Fannin, Lamar, and Delta Counties the Pecan Gap Chalk is consistently 50 feet of bluish-gray, slightly bituminous, argillaceous to sandy chalk. It weathers light gray to white (see Plate 1-Fig.s 8 and 9). Megafossils are moderately common and scattered throughout.. The lower contact with the Wolfe City Sandstone is unconformable and characterized by a condensed section. At the type locality the basal chalk is glauconitic, rich in phosphate nodules and phosphate moulds of molluscs, especially *Baculites*. Chalk-filled borings extend into the Wolfe City Sandstone. This basal zone is easily identified in the field as the phosphatic material weathers out of the chalk matrix and is concentrated by the weathering process. The upper contact with the Marlbrook Marl is conformable.

PALEOGEOGRAPHY of the NSR STUDY AREA

As stated earlier, Fig. 6 is a paleogeographic reconstruction of the United States at the base of the Maastrichtian stage 70 million years ago. It has been compiled from numerous lines of evidence including macro and microfossil compilations, formational facies determinations, structural geology relationships, and many others. Paleogeographic reconstructions are the result of years of research by academia, private oil companies, research consortiums, and various government agencies. There is even a commercial product on the market now that rotates all the various crustal plates of the Earth backwards through time to the base of the Cambrian and beyond.

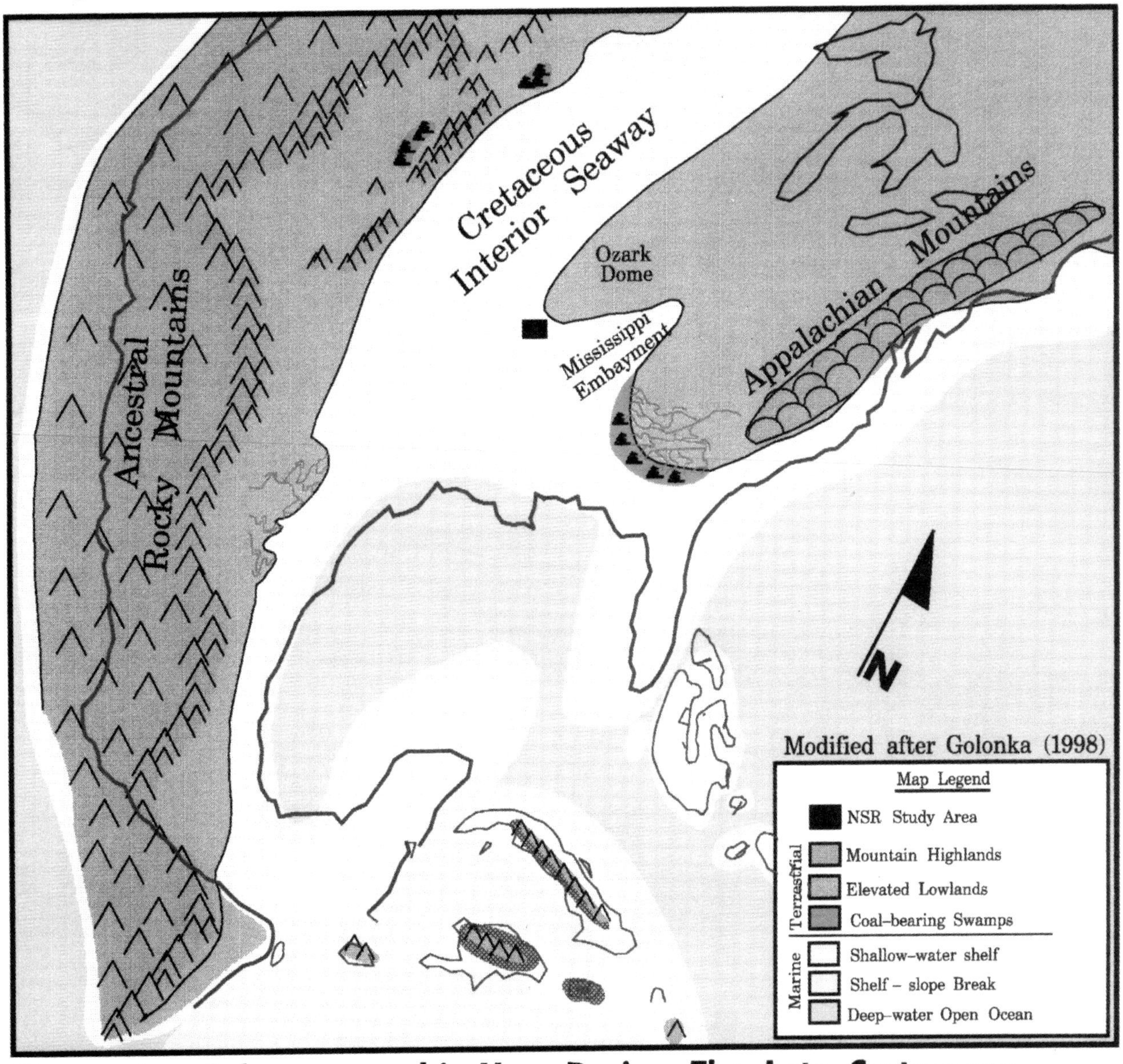

Figure 6. - Paleogeographic Map During The Late Cretaceous (Maastrichtian Stage - 70 MYA)

As can be seen in Fig. 6, the Western Interior Seaway extended from the Gulf of Mexico northward and essentially split the present-day United States in half. The present-day outline of North America is included on the map to give a perspective of how much of the continent was underwater during the Late Cretaceous. The NSR study area is highlighted by the solid black rectangle. It was located at the extreme southeastern corner of the Western Interior Seaway. The then much larger Gulf of Mexico opened up to the south. An arm of the Gulf of Mexico extended up the present day Mississippi River Valley and is called the Mississippi Embayment.

All of Texas was underwater during the Campanian except for the extreme western portion of the state. Active mountain building and uplift kept this area from the Davis Mountains westward above sea level. Also, the shoreline in north Texas was very close to the present-day Red River and migrated back and forth across the northern counties with minor sea level transgressions and regressions. The Western Interior Seaway was essentially an asymmetrical trough with the steeper slope on the western margin facing the ancestral Rockies. The more gently sloping eastern margin lapped up against the midcontinent craton. Water depths in the Western Interior Seaway were never very deep. It has been estimated that the deepest part of the seaway was centered in the present-day Colorado Plateau area. Water depths may have reached down to 1000 feet (300 meters). The Pierre Shale in western Kansas was deposited in water depths no greater than 330 feet (100 meters).

Paleontologic and lithologic evidence indicates that the NSR Campanian sediments were deposited over a broad shelf with topographic highs to the west and to the northeast. Sediment was supplied by a source to the northwest. This was probably the exposed core of the Quachita Foldbelt in present-day Oklahoma. Microfossil and sediment distribution indicates that surface currents swept from west (out of the Western Interior Seaway) to east towards the Mississippi Embayment.

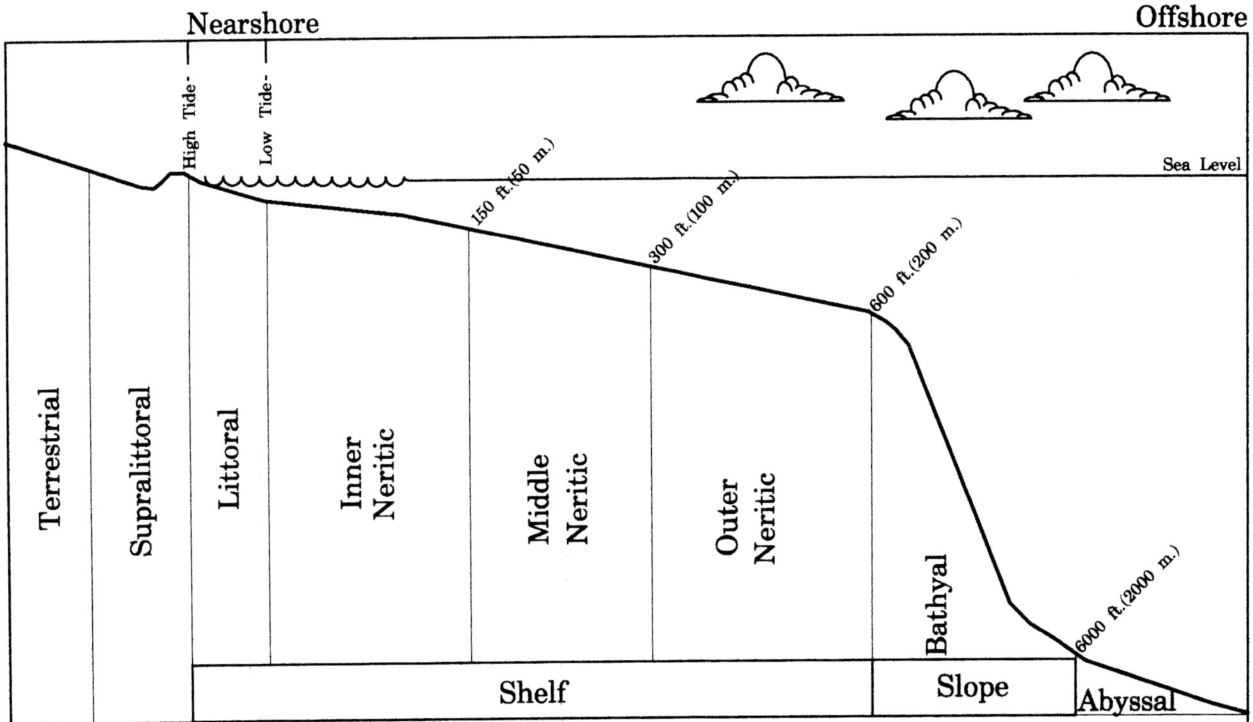

Figure 7. – Classification of Marine Environments
Based on depth of water column to sea floor.

Fig. 7 is a diagram showing the classification of marine environments based upon water depths. When you hear the term "continental shelf" in the news, they are referring to water depths of 600 feet or less. This roughly corresponds to the shelf-slope break in the Gulf of Mexico. If you were to take a boat due south from Houston, Texas, you would have to travel 116 miles to reach the shelf-slope break. All the seafloor in between would be considered part of the continental shelf. And yet in some places in offshore California you can reach the 600-foot contour within a mile from the coastline! It varies from place to place along all the coasts across all the continents in the world. The Pierre Shale in western Kansas at 330 feet would be considered as having been deposited in the middle neritic zone.

In the oceans of the world today, there are countless forms of benthic foraminifera. They prefer certain ranges of temperature, light, nutrient levels, oxygen levels, current velocities, etc.. Therefore the different forams tend to concentrate on the sea floor at certain depth intervals based upon the conditions they favor. In addition, there are countless forms of planktonic forams living at various depths in the water column that also have environmental preferences. Over the years, marine scientists have sampled the continental shelves and have been able to chart which types of foraminifera inhabit what depth ranges.

These forams can then be used as depth indicators in the geologic past. Oil companies employ biostratigraphers to chart occurrences of forams and other microfossils from well cuttings. Well cuttings are the pieces of bedrock carried to the surface while an oil or gas well is being drilled. Just like rock formations exposed on the surface, they commonly contain fossils. Biostratigraphers use the microfossils to determine paleoenvironments of the reservoirs containing oil and gas deposits. Over the years an extensive database of foraminifera and other marine microfossils has been established that can be used to determine marine environments of formations containing those fossils.

Extensive research has been done on the forams and other microfossils of the various formations in the Taylor Group of northeast Texas. These formations were deposited in shallower water than the Pierre Shale mentioned above as they were much nearer to the paleoshoreline. Heine (1991, p.117) used cluster analysis on dinoflagellate cysts (microflora) to conclude that the Ozan Formation was deposited in a middle neritic shelf environment. Water depths ranged from 150-300 feet. Population and diversity plots indicated turbid conditions on the seafloor with abundant, clay-sized terrigeneous material. The sea bottom would have been muck with lots of clay-sized particles.

At its type locality, Selznick (1991, p.77) showed that the Wolfe City Sandstone had an extremely low diversity foraminiferal fauna. In conjuction with its coarser-clastic lithology (it is the only sand facies within the Taylor Group), and the fact that it is dominated by a shallow-water molluscan assemblage, it is safe to assume that the Wolfe City Sandstone was deposited in the shallowest water environment of any formation within the Taylor Group. It was probably deposited in the inner neritic zone in water depths of 150 feet or less. Sand studies indicate that the source of the clastic material was to the north/northwest and possibly as close as 30 miles away. Most likely the Wolfe City Sandstone was part of a delta complex that climaxed a minor sea level regression. Sand deposition implies a certain amount of current activity, which indicates the sea bottom would have been well oxygenated and probably nutrient-rich. This helps explain the abundant molluscan fauna and the ichnofossils found in the sandstone.

References

Akers, R.E. and Akers, T.J. 1987. Texas Cretaceous Echinoids. Paleontology Section, Houston Gem and Mineral Society, 3:144 p.

Akers, R.E. and Akers, T.J. 1997. Texas Cretaceous Gastropods. Paleontology Section, Houston Gem and Mineral Society, 6:340 p.

Barrow, T.D.,1953, Mesozoic stratigraphy and structural history of the East Texas Basin. Stanford University unpublished phD dissertation, 221 p.

Bottjner, J.D. 1979. Ichnology and depositional environments of Upper Cretaceous chalks, Southwest Arkansas (Annona Fm, Chalk Mbr. Saratoga Fm). Bulletin of the American Association of Petroleum Geologists, 63(4):422.

Bottjner, J.D. and Bryant,W.A. 1980. Paleoenvironmental analysis of disconformity and condensed bed at contact of Austin and Taylor Groups (Upper Cretaceous), East- Central and Northeastern Texas. Bulletin of the American Association of Petroleum Geologists, 64(5):679.

Bottjner, J.D. 1986. Campanian – Maastrichtian chalks of Southwestern Arkansas: petrology, paleoenvironments, and comparison with other North American and European chalks. Cretaceous Research, 7:161-196.

Cobban, W.A. and Kennedy, W.J. 1992. Campanian ammonites from the Upper Cretaceous Gober Chalk of Lamar County,Texas. Journal of Paleontology, 66(3):440-454.

Cobban, W.A. 1992. Campanian *Trachyscaphites spiniger* ammonite fauna in Northeast Texas. Paleontology, 35(1):63-93.

Cobban, W.A. and Kennedy, W.P. 1993. Middle Campanian ammonites and inoceramids from the Wolfe City Sand in northeastern Texas. Journal of Paleontology, 67(1):71-82.

Dane, C.H., 1929, Upper Cretaceous formations of southwestern Arkansas. Arkansas Geological Survey, 1:215 p.

Echols, J. 1984. Ammonite zonation in condensed zone, Middle Ozan Formation (Taylor Group – Upper Cretaceous) in northeast Texas. Bulletin of the American Association of Petroleum Geologists, 68(4):473.

Emerson, B.L., Emerson, J.H., Akers, R.E., and Akers, T.J. 1994. Texas Cretaceous Ammonites and Nautiloids. Paleontology Section, Houston Gem and Mineral Society, 5:440 p.

Finsley, C. 1989. A field guide to Fossils of Texas. Texas Monthly Press, 190 p.

Hicks, J.F., Obradovich, J.D., and Tauxe, L. 1998. Magnetostratigraphy, isotropic age calibration and intercontinental correlation of the Red Bird section of the Pierre Shale, Niobrara County, Wyoming, USA. Cretaceous Research, 20: 1-27.

Hill, R.T. 1894. Geology of parts of Texas, Indian Territory, and Arkansas adjacent to the Red River. Geologic Society of America Bulletin, 5:297-338.

Kauffman, E.G. 1977. Illustrated guide to biostratigraphically important Cretaceous macro-fossils, Western Interior Basin,USA. Mountain Geologist, 14:225-274.

Kennedy, W.J. and Cobban, W.A. 1991. Upper Cretaceous (upper Santonian) *Boehmoceras* fauna from the Gulf Coast region of the United States. Geological Magazine, 128(2): 167-189.

Kennedy, W.J., Cobban, W.A. and Scott,G.R. 1992. Ammonite correlation of the uppermost Campanian of western Europe, the U.S. Gulf Coast Atlantic Seaboard and Western Interior, and the numerical age of the base of the Maastrichtian. Geology, 129(4):497-500.

Kennedy, W.J. and Cobban, W.A. 1993. Campanian ammonites from the Annona Chalk near Yancy, Arkansas. Journal of Paleontology, 67(1):83-97.

Kennedy, W.J., Cobban, W.A. and Landman, N.H. 1997. Campanian ammonites from the Tombigbee Sand Member of the Eutaw Formation, the Mooreville Formation, and the basal part of the Demopolis Formation in Mississippi and Alabama. American Museum Novitates, 3201: 1-44.

Larson, N.L., Jorgensen, S.D., Farrar, R.A. and Larson, P.L. 1997. Ammonites and the other cephalopods of the Pierre Seaway. Geoscience Press,Inc., 148 p.

McNulty, C.L., Jr. 1977. Emandation of the Pecan Gap Chalk (Campanian) of northeastern Texas. Transactions of the Gulf Coast Association of Geological Society, 27:430.

McNulty, C.L., Jr., Brezina, J.L., Dawson, W.C. and Malv, F.W. 1981. Emandation of the Pecan Gap Chalk (Campanian) of northeastern Texas. Transactions of the Gulf Coast Association of Geological Society, 31:353-358.

Nicholls, E.L. and Russell, A.P. 1990. Paleobiogeography of the Cretaceous Western Interior Seaway of North America: the vertebrate evidence. Paleogeography, Paleoclimatology, PaleoEcology, 79:149-179.

Pate, K.A. 1996. The significance of the fossil communities of the Wolfe City Sand of the Taylor Group (Upper Cretaceous, Gulfian Series), central Texas. Bulletin of the Geological Society of America, 28(1):58.

Pessagno, E.A.,Jr. 1969. Upper Cretaceous stratigraphy of the western Gulf Coast area of Mexico, Texas, and Arkansas. Geological Society of America Memoir, 111:139 p.

Richardson, J.D. 1972b. Paleoecology of ichnofossils from the Wolfe City Formation (Upper Cretaceous), northeast Texas. Geological Society of America Abstracts with Programs, 4(4): 290-291.

Rouse, J.T. 1944. Correlation of the Pecan Gap, Wolfe City, and Annona Formations in east Texas. Bulletin of the American Association of Petroleum Geologists, 28(4):522-530.

Stehli, F.G., Creath, W.B., Upshaw, C.F. and Forgotson, J.M. 1972. Depositional history of Gulfian Cretaceous of the East Texas Embayment. Bulletin of the American Association of Petroleum Geologists, 56(1):38-67.

Stephenson, L.W. 1918. A contribution to the geology of northeastern Texas and southeastern Oklahoma. United States Geological Survey Professional Paper, 120-H:129-163.

Stephenson, L.W. 1927. Notes on the stratigraphy of the Upper Cretaceous formations of Texas and Arkansas. Bulletin of the American Association of Petroleum Geologists, 2(1):1-17.

Thompson, L.B., Percival, S.F.,Jr. and Patricelli, J.A. 1978. Stratigraphic relations of the Annona Chalk and Gober Chalk (Upper Campanian); type localities in northeast Texas and southwest Arkansas. Transactions of the Gulf Coast Association of Geological Society, 28:665-679.

Thompson, L.B., Heine, C.J., Percival, S.F.,Jr. and Selznick, M.R. 1991. Stratigraphy and micropaleontology of the Campanian shelf in northeast Texas. Micropaleontology Special Publication, 5:148 p.

Tucker, D.R. and Hencey, D.P. 1987. Mini-shelves in Taylor shale: stratigraphy of Late Cretaceous and Early Tertiary units, Bastrop County, Texas. Bulletin of the American Association of Petroleum Geologists, 71(9):1123.

University of Texas – Bureau of Economic Geology. 1966. Geologic Atlas of Texas – Texarkana sheet. University of Texas – Bureau of Economic Geology, scale 1:250,000.

Welton, B.J. and Farish, R.F. 1993. The collector's guide to fossil sharks and rays from the Cretaceous of Texas. BeforeTIME Publishing, 204 p.

Young, K.P. 1963. Upper Cretaceous ammonites from the Gulf Coast of the United States. University of Texas Publication, 6304:373 p.

Chapter 2 Cretaceous Invertebrates

The NSR contains an abundant and diverse invertebrate fossil assemblage. Macrofossils were recorded by Hill (1894), followed by Stephenson (1918), and Dane (1929) in the early part of the last century. These faunal lists formed the basis for species identification for the NSR for the next fifty years. In the last decade, attention to the molluscan assemblage, especially the ammonites, has resulted in a series of papers by Cobban & Kennedy (1992,1993) on the faunas of the Gober Chalk, Ozan Formation, and the Wolfe City Sandstone in the NSR area. The paleontology section of the Houston Gem & Mineral Society has done a series of papers on various Cretaceous invertebrate groups (1987,1994,1997) that included fossils from the NSR. In addition, an excellent summary of the microfossil work done for the entire Taylor Group was compiled by Thompson et al (1991).

Tables 1, 2, and 3 are macrofossil listings for the Ozan, Wolfe City, and Pecan Gap Formations, respectively. These have been compiled from all available historical and recent faunal listings. There is no way to verify the accuracy of some of the older published faunal lists without seeing the collections they were based on. Some fossil names change over the years as they are restudied. Genera are renamed because an older fossil name with priority is applied to it. In addition, the tables incorporate the results of fossils brought into the NSR Symposium. Even after all these years of study there were still plenty of new additions to the faunal lists.

For the Ozan Formation, we recorded 12 previously unreported invertebrates from the NSR. These included three shrimp/lobsters (all unidentified), the brachiopod *Choristothyris*, an unidentified bryozoan, three carophyllid corals, two unidentified starfish, a articulate crinoid and an unidentified cidarid echinoid. From the Pecan Gap Chalk the calcareous sponge *Cliona* (Plate 2, Fig. 10 and 11) and the scleractinian corals *Trochocyathus* (Plate 2, Fig. 7), and *Cladophyllia/Caulastrea*? are recorded for the first time. Also recovered from the Pecan Gap Chalk are the oral plates from the barnacle *Acroscalpellum sp.* (Plate 2, Fig. 5) and the twig-like bryozoan *Coscinopluera*. No new fossil genera were reported from the Wolfe City Sandstone. Most of the reported macrofossils from the Wolfe City come from the type locality just northeast of the town of Wolfe City. A total of 17 new invertebrate genera were recorded from the NSR.

It was interesting to see the number of really good invertebrate fossils that were better specimens than those illustrated in previous guidebooks and publications. With so much attention focused on the vertebrate remains from the NSR, the invertebrates were frequently overlooked. So much for assumptions! Collectors were just waiting for a forum to showcase their treasures. Tables 1 through 3 give no indication as to the scarcity or abundance of any given fossil. For example, in Table 1 most of the fossils from the Ozan Formation outside of the molluscs (including the cephalopods, gastropods, and pelecypods) are represented by single occurrences. The most ubiquitous invertebrate fossil from the NSR is the large oyster *Exogyra ponderosa*. It is found along the length of the NSR in all formations. With its large size and characteristic shape (Plate 5, Fig. 7 and 8), it is easily spotted in the gravel bars from great distances.

The Ozan Formation has an abundant and diverse invertebrate macrofossil assemblage. Most of the major phyla are represented. The marl lithology indicates a muddy, soft substrate that would not be suitable for reef-building organisms. Hence, the paucity of corals, bryozoans, sponges or other reef builders. These organisms require a firm substrate for initial colonization and the Ozan seafloor would have been too soupy for them to settle on. They also require relatively clear water. The Ozan sea floor interface had a high content of clay-sized sediment in suspension. The two bryozoan specimens brought in were from very small twig-like colonies. Both specimens were under 5 mm in length. The solitary corals are also very small forms that are easily overlooked while surface collecting.

Another potential reef-builder is the rudist clams. Rudists are represented on the NSR by *Durania* and *Sauvagesia* (Plate 5, Fig. 9-11). This strange group of bivalves is characterized by very large, thick, non-equal sized valves. Its lower valve was elongated and grew in the conical form of "horn corals" with its distal apical end buried in the sea floor as in Fig. 1. This valve was generally very thick (greater than four inches in larger mature specimens of *Durania*) and had a longitudinally fluted exterior surface. Its upper valve faced upwards, was lid-like and acted like an operculum. They grew individually or in colonies that sometimes fused together into "patch" reefs. Rudist reefs are common in the lower Cretaceous Edwards Formation south of the DFW area. They first appeared in the upper Jurassic and went extinct towards the end of the Maastrichtian. Most collectors mistake rudist clams for something else. Their weathered internal structure can be misidentified as fossil wood.

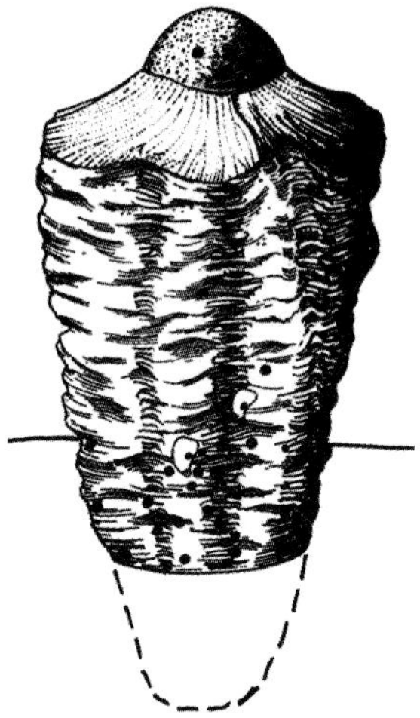

Fig. 1 Reconstruction of *Durania austinensis*. The dashed lnes indicate the part of the valve buried in the sediment. Rudists resemble horn corals in their overall shape and life position.

The Pecan Gap Chalk represented bottom conditions more conducive to the formation of localized hardgrounds with reef building type organisms. There is an increase in rudistids in the chalk versus the marly Ozan Formation or the sandy Wolfe City Sandstone. However, there are still no organic buildups that can be considered true reefs.

Exogyra ponderosa probably formed extensive oyster banks along the sea floor. While not true reefs, they were local topographic highs that disrupted bottom currents and probably acted as magnets for other invertebrates to colonize around. Their valves are also very thick and different in size and shape. Most oysters, including *Ostrea* and *Anomia* from the NSR, attach to hard surfaces. This is called a "fixed-sessile" lifestyle. Fixed because they are attached to something and sessile because they don't move after settling on an object for the rest of their adult lifetime. These oysters have the ability to change their shape as they grow to match the substrate they are on. This also means they have a great variability in shell form. There are numerous different species of *Ostrea* from the Cretaceous of North America. Many of the species are probably invalid and represent shell variations within single populations described from one locality.

The only other common large bivalve is the flatter *Inoceramus*. All other pelecypods arecommonly preserved as phosphatized internal molds which makes species identification difficult. *Cardium* is a burrowing clam that lives under the sediment water interface. It has two siphons, one inhalant for gas and food capture and one exhalant for waste disposal. Swimming forms present in the NSR include *Pecten*. It can rise from the sea floor and swim for several meters by rapidly opening and closing its valves.

The cephalopods are the most highly evolved group of molluscs as well as one of the most highly evolved marine invertebrates period. Living representatives include *Octopus* (octopus), *Sepia* (squid), and the *Natilius* (nautiloid) among others. The *Natilius*, with its coiled external shell, is the closest analog we have for the extinct ammonites. The coiled nautiloids reached their evolutionary apex during the Paleozoic and were rapidly displaced by the ammonites during the Mesozoic. The *Natilius* of today is a swimming/free-floating animal that is nocturnal by nature. It spends the daylight hours at great depths in the darkness for protection. It comes to the surface at night to feed and is an active carnivore. Besides the shell, the only other part of an ammonite that is commonly preserved is the radula. This is its jaws and they look like a parrot beak. They are commonly mistaken for pelecypod valves if they are not found in place within the living chamber of an ammonite.

A majority of the well preserved cephalopods of the Ozan Formation come from one locality on the NSR. It is located one and a quarter miles upstream from the FM 2990 bridge north of Ladonia. Just inside the mouth of Davis Creek (on the south side of the river) is an exposure of the informal "red zone" along the creek floor and its banks (Plate 1, Fig. 5-7). This is the single richest invertebrate layer in the whole Ozan interval. It is also exposed just upstream from the creek entrance on both sides of the river. The layer is a red-weathering glauconitic marl loaded with bioclasts. It contains both phosphatized molds of *Baculites* and other molluscs, and ammonites and gastropods with their original aragonitic shell material. Cobban and Kennedy (1992) date this zone at approximately 80 millions of years ago (MYA) based on radiometric dates from Western Interior bentonite beds.

The gastropods are also well represented from the NSR. There are two orders of marine gastropods, the Prosobranchia and the Opisthobranchia. All species of prosobranchs have calcareous shells that can be fossilized. The most common mode of preservation is phosphatized internal molds found in the gravel banks. Gastropods are a highly successful group of marine molluscs today. They occupy all types of substrates and are carnivores, herbivores, and scavengers. There is no relation to shell shape and lifestyle. You cannot tell by a high-spired or low-spired shell if it lived on a sandy bottom and ate algae or burrowed in the muck.

The most common form is *Gyrodes major* (Plate 6, Fig. 1). In the "red zone" in the Ozan Formation, gastropods are found with their original shell material and ornamentation intact. There turns out to be quite a wide variety of gastropods as indicated in Table 1. Keep in mind that some of the older published faunal lists were based on internal gastropod molds only and probably have some misidentifications.

The fossil shrimp/lobsters were a pleasant surprise. There is a literature reference to fossil decapods from the Roxton Limestone Member of the Gober Chalk. These were from an abandoned road quarry in the Roxton outcrop belt north of the town of Roxton in Lamar County. No crustaceans had previously been reported from the NSR itself. The order Decapoda is a well defined group of the Malacostraca or "higher crustaceans". They are characterized by having three pairs of maxillar limbs and five pairs of thoracic limbs which are called pereiopods. The carapace covers the entire head and the abdomen. Shrimp, crabs, and lobsters belong to the suborder Natantia and are walkers and crawlers. Two of three different genera are small phosphatized forms (Plate 3, Fig. 1 and 2) and the third is a much larger tail and abdomen section resembling *Linuparus* (Plate 3, Fig. 3). *Linuparus* and some closely related forms are long ranging genera lasting throughout the late Cretaceous. No identifiable crabs were brought into the NSR Symposium though a possible very large carapace is illustrated (Plate 3, Fig. 4 and 5).

Echinoderm macrofossils from the Ozan Formation were represented by asteroids (starfish), crinoids (sea lillies) and echinoids (sea urchins). Two different kinds of undescribed asteroids were brought to the seminar. One type, which is represented by arm fragments only (Plate 3, Fig. 6) appears to belong to the Family Astropectinidae. *Astropecten* is a living fossil and its skeleton is composed of large and heavy plates. The other (Plate 3, Fig. 7 and 8) is an enrolled specimen that is nearly complete but in four associated pieces.

On a DPS fieldtrip years ago, Mike Murphy from Ardmore, Oklahoma collected a slab with multiple complete crinoids on it from a tributary feeding into the NSR. The whole crinoid colony appeared to be attached to a floating mass of seaweed (kelp?) or a piece of driftwood. The colony died when the floating attachment site sank to the bottom. Only the crinoid remains were preserved. The slab is currently being described and is not available for illustration.

Crinoid stem and root fragments were recovered from the Highway 38 bridge at Ben Franklin downstream past the Highway 24 bridge. They are very small, slender and easily overlooked. They can be identified to the family level as belonging to the Bourgueticrinidae or the Bathycrinidae. This type of articulate crinoid is characterized by a very small fusiform to pyriform cup on a long slender stem of variable shape and attached to a well developed root system. The calyx is small and appears more as a swelling at the proximal end of the stem. The most likely genus is *Bourgueticrinus* or *Dunnicrinus. Bourgueticrinus* has been reported from the Maastrichtian of Alabama and *Dunnicrinus* from the Maastrichtian of Mississippi in the same type of marlstones as found along the NSR. Careful searching should produce better specimens.

Specimens of the heart urchin *Hemiaster* were brought in from all of the formations in the NSR study area (Plate 4, Fig. 1-4). It apparently ranged throughout the mid Campanian in northeast Texas though it never occurred in great numbers based on preserved body fossils. However, a lot of the heavily bioturbated rock beds from the early Cretaceous chalks of north Texas are attributed to the infaunal burrowing activity of heart urchins. They may have been much more common in the NSR chalk beds than fossilized remains indicate.

Echinoids are a class subdivision of the phylum Echinodermata. They are further subdivided into two subclasses based on their body organization - regular and irregular. Heart urchins are familiar fossils in the early Cretaceous of north central Texas. The spatangoids (heart urchins) lived under the sediment-water interface in burrows. They have lost their original five-fold symmetry as an adaptation to their infaunal lifestyle. Notice in Plate 4, Fig. 1 and 2 how the overall shape of *Hemiaster* resembles a heart. Their mouths have migrated from a central position on the underside of the test to an anterior position. They filtered out organic detrius as they ingested sediment under the sea floor. They were covered in very small and slender spines that are very rarely preserved attached to the test. If you have ever dug up living sand dollars on a sandy beach at low tide then you can imagine the life habit of a heart urchin. Its test is very thin and weak and crushes easily. Complete heart urchin fossils indicate that the animal remained buried after death long enough for the sediment to solidify around it.

On the other hand, regular echinoids like the modern cidarids (Plate 4, Fig. 5, 8 and 9) live on the sea floor and are active carnivores. They retain their five-fold (pentameral) symmetry. Cidarid echinoids have a greater variety and larger spines than do the heart urchins. Some feed on algae that they scrape off of hard surfaces. Regular sea urchin mouthparts are much different than those in heart urchins. The associated dentitions are known as "aristotle's lantern" but are rarely preserved as fossils. Regular echinoids overall have less chance of being preserved as complete tests than heart urchins due to the fact that they lived on rather than under the sea floor. There is a much greater chance of being destroyed by mechanical means (wave action) or being consumed (fish, cephalopods, etc.) if you expire on the sea floor versus dying while buried under a protective layer of sediment. In some parts of the world regular urchins are areal problem as armies of them are destroying coral reefs.

One fossil organism that must have been extremely common but is very rarely preserved as a body fossil is the calcareous sponge *Cliona cretacia* (Plate 2, Fig. 10 and 11). Evidence of its predation on pelecypod shells can be seen in the form of small round holes bored into *Exogyra ponderosa* valves. Specimens of *Exogyra ponderosa* shells from the entire length of the NSR commonly have these borings in them.

There are many kinds of plant and invertebrate fossils that are very rarely preserved in the fossil record. These include algae, seaweeds, seagrass, jellyfish, sea slugs, sea anemones, polycheate worms, and others. This does not mean they were not common constituents of the benthic community. In fact, some of them may have been the most abundant elements of the biota. It is just that in lacking hard parts, they left no record of their passing.

Other large organisms can only be found by looking for the microscopic evidence. For example, holothurians (or sea cucumbers) can get over a foot in length but the only fossils they leave behind are tiny micoscopic sclerites. Sclerites are calcareous skeletal elements of various shapes that supported the body but did not form a continuous rigid network. Sponges may also have been a lot more common during the mid Campanian in northeast Texas. Their bodies are composed of a mesh of calcareous or opaline(siliceous) spicules. These can be common microscopic fossils even though complete individuals are rare.

So far, we have only discussed what are considered to be body fossils. These are the actual remains or molds of the organisms themselves. There is another branch of paleontology that investigates the traces left by the life activities of the fossil organisms. This is the field of paleoichnology or the study of trace (ichno) fossils. Trace fossils can be caused by locomotion activity (arthropod tracks), feeding strategies (grazing trails), excretion of body wastes (coprolites), or interactions with other organisms (sponge boreholes in oyster valves).

Table 2 lists some ichnofossils from the Wolfe City Sandstone. All of these are attributed to various crustaceans. For example, *Thalassinoides* are the branching, dwelling burrows of *Callianassa* or *Upogebia* (both are shrimp with living representatives today). These were the mucus-lined tunnels that the decapods retreated into when danger approached or they were resting. Even though the body fossils have not been positively identified from the NSR yet, we can assume that they (or their close cousins) were present based on the trace fossils they left behind. Those of you who have collected the Britton Formation "crab zone" at the Lake Lewisville Spillway in Denton County, Texas are familiar with these burrows. The burrows at this locality are so abundant and the activity so great that they have coalesced into sheets along some horizons. Sediment infilling and later mineralization produced the red nodule zones seen there today.

Fig. 2 Reconstructed seascape based on selected faunal elements from the Late Cretaceous Ozan Formation.

Fig. 2 represents a possible seascape from 80 million years ago during deposition of the Ozan Formation in northeast Texas. Vertebrate faunal elements (from top to bottom) include the mosasaur *Platecarpus ictericus* pursuing a school of fleeing fish. The large ichthyodectid fish *Xiphactinus audax* lies motionless below the marine reptile waiting for an unsuspecting meal to swim by. Another predatory fish, *Pachyrhizodus minimus*, swims past the large ammonite *Pachydiscus paulsoni*. Meanwhile, an unidentified pycnodont fish is busy scraping algae off a large demosponge colony. Oblivious to it all is the large marine sea turtle *Protostega* who propels itself through the waters with its mighty paddles.

Ammonites are numerous and varied during the Campanian. The heteromorph *Baculites cf. aquilaensis* seems to hang suspended upside down. Buoyancy studies have shown that its center of gravity is such that it floated vertically in the water column. It is still debated which way the animal itself faced (up or down). Modern day squids are gregarious creatures and are encountered in groups of hundreds to thousands. Scientists assume baculitid ammonites were also social animals. Another heteromorph, *Glyptoxoceras sp.* is seen approaching a cluster of *Inoceramus balticus* clams.

In the left foreground are two large *Exogyra ponderosa* oysters. The nearest one is encrusted with the serpulid worm *Hamulus onyx*. A colony of the boring sponge *Cliona cretacia* has also settled on it. The sponge will eventually kill its host as it bores through the thick shell to reach the inner mantle cavity. Other molluscs include the gastropod *Gyrodes major* inching along the seafloor and the gastropod *Pluerotomaria sp.* crawling up the demosponge colony.

Echinoderms are represented by the heart urchin *Echinocorys texanus* which lies partially buried in the seafloor. Nearby rests an unidentified starfish. Other benthic invertebrates include the scleractinian coral *Trochocyathus sp.* Not shown are the numerous varieties of seaweed and seagrass that possibly carpeted the seafloor at this time.

Always keep in mind that the fossils you collect only tell us part of the story. We will never get a complete picture of what strange creatures inhabited the mid Campanian in northeast Texas. Preservational factors, depositional processes and collecting bias hinder the end result. This is one reason the NSR Symposium and the publication of this guidebook is so important. It helps to fill in the knowledge gaps and give us a more complete picture of past life along the NSR.

Table 1. Macrofossil Faunal List

AGE: Lower Mid Campanian Stage/Gulfian Series/Upper Cretaceous
FORMATION: Ozan Formation/lower Taylor Group
LOCATION: North Sulphur River, northeast Texas

PHYLUM ANNELIDA
Hamulus squasmosus
Hamulus onyx
Proliserpula (Serpula) sp.

PHYLUM ARTHROPODA
CLASS CRUSTACEA
ORDER DECAPODA
SUBORDER ASTACIDEA
Enoploclytia kinzeyi?
Hoploparia sp.
Linuparus sp.

PHYLUM BRACHIOPODA
CLASS ARTICULATA
ORDER TERBRATULIDA
Choristothyris plicata

PHYLUM BRYOZOA
Unidentified slender, branch-like colony

PHYLUM COELENTERATA
CLASS ANTHOZOA
ORDER SCLERACTINIA
SUBORDER CAROPHYLLINA
Trochocyathus taylorensis (solitary trochoform form)
Astrocoencia sp. (colonial form)
Micrabacia hilgardi ("psuedo"colonial form)
Micrabacia roltilia (solitary "button"coral)

PHYLUM ECHINODERMATA
CLASS ASTEROIDEA
ORDER
FAMILY ASTROPECTINIDAE
Specimen #1 - unidentified arm fragments
Specimen #2 – enrolled (family unknown)

CLASS ECHINOIDEA
ORDER SPATANGOIDA
Echinocorys cf. Texana
Hemiaster beecheri?

ORDER CIDAROIDA
Unidentified cidarid

PHYLUM MOLLUSCA

CLASS CEPHALOPODA

ORDER AMMONOIDEA

Baculites sp.cfr.aquilaensis
Glyptoxoceras sp.
Menabites (D.) delawarensis
Pachydiscus paulsoni
Placenticeras sp.
Submortoniceras aff. *condamyi*
Trachyscaphites spiniger spiniger
Eupachydiscus grossouvrei
Menabites(Delawarella)danei
Menabites (D.) vanuxemi
Patagoisites sp.
Scaphites hippocrepis
Trachyscaphites densicostatus

ORDER NAUTILOIDEA

Cinomia ozani
Eutrephoceras dekayi

CLASS GASTROPODA

Anchura substriata
Anisomyon borealis
Gyrodes major
Pyropsis proxima
Volutomorpha aff. *conradi*
Actaeon sp.
Delphinula cf D. lapidosa
Natica sp.
Turritella trilira
Volutederma sp.

CLASS PELECYPODA

Anisomyon borealis
Cardium sp.
Cymbophora sp.
Durania (Radiolites) austinensis [rudist]
Gastrochaena americana
Inoceramus (Endocostea) balticus
Lima aff. reticulata
Limposis cf. meeki
Paranomia scabra
Pecten aff. quinquecostatus
Ostrea panda
Sauvagesia cf. Belti [rudist]
Veniella conradi
Anomia argentaria
Crassatellites sp.
Cyprimeria sp.
Exogyra ponderosa
Gryphaeostrea sp.
Inoceramus barabini
Liopistha alternata
Nuculana perequalis
Pecten hilgardi
Ostrea falcata
Ostrea plumosa
Texigryphaea vesicularis

PHYLUM CHORDATA

CLASS CHONDRICHTHYES

SHARKS

Cretolamna appendiculata
Ginglymostoma lehneri
Hexanchus microdon
Lissodus sp.
Pararhincodon groessenssi
Psuedocorax granti
Ptychodus lattissimus
Serratolamna serrata
Squalicorax pristodontus
Squatina hassei
Galeorhinus sp.
Heterodontus canaliculatus
Lissodus selachos
Palaeogaleous sp.
Paranomotodon sp.
Ptychodus connellyi
Scapanorhynus texanus
Squalicorax kaupi
Squalus sp.

RAYS

Brachyrhizodus wichitaensis
Ischyrhiza avonicola
Onchosaurus pharao
Ptychotrygon triangularis
Sclerorhynchus sp.
Dasyatis sp.
Myliobatis ps.
Ptychotrygon agujaensis
Rhinobatos casieri

CLASS OSTEICHTHYES

ORDER AULOPIFORMES

Enchodus ferox

ORDER ELOPIFORMES

Pachyrhizodus minimus
Pachyrhizodus caninus

ORDER ICHTHYODECTIFORMES

Gillicus arcuatus
Saurodon sp.
Xiphactinus audax

ORDER PACHYCORMIFORMES

Protosphyraena sp.

ORDER PYCNODONTIFORMES

Anomoeodus barberi

ORDER TETRAODONTIFORMES

Stephanodus sp.

CLASS REPTILIA

ORDER TESTUDINES

Unidentified turtle scutes

ORDER CROCODILIA

Unidentified dyrosaurid? scutes

ORDER SQUAMATA

Clidastes propython
Globidens alabamaensis
Halisaurus sp.
Platecarpus tympaniticus
Tylosaurus proriger

SUBCLASS DINOSAURIA

ORDER ORNITHISCIA

FAMILY HADROSAURIDAE

Unidentified hadrosaur tooth plate

Table 2. Macrofossil Faunal List

AGE: Middle Campanian Stage/Gulfian Series/Upper Cretaceous
FORMATION: Wolfe City Sandstone/lower Taylor Group
LOCATION: North Sulphur River, northeast Texas

TRACE FOSSILS

Chomatichnus
Gyrolithes
Rhizophycus
Cruziana
Psuedobilobites
Thalassinoides

(Note: all of these are indicative of a crustacean-rich epifauna)

PHYLUM ANNELIDA

Hamulus squasmosus
Hamulus onyx
Proliserpula (Serpula) sp.

PHYLUM COELENTERATA

CLASS ANTHOZOA

ORDER SCLERACTINIA

Hindesastracea collinensis (colonial form)
Trochocyathus taylorensis (solitary form)

PHYLUM MOLLUSCA

CLASS CEPHALOPODA

ORDER AMMONOIDEA

Baculites asper
Hoplitoplacenticeras marroti
Patagoisites sp.
Trachyscaphites spiniger porchi
Baculites mclearni
Hoplitoplacenticeras minor
Placenticeras sp.

ORDER NAUTILOIDEA

Eutrephoceras dekayi

CLASS GASTROPODA

Acmaea sp.
Acteon sp.
Anisomyon sp.
Beretra sp.
Calliomphalus americanus
Ecphora proquadricosta
Graciliala calcaris
Gyrodes major
Laxispira lumbricalis
Lomirosa cretacea
Morea marylandica marylandica
Paladmete gardnerae gardnerae
Pugnellus densatus
Anchura substriata
Acteon texana
Bellifusus sp.
Buccinopsis crassa
Cantharus sp.
Euspira rectilabrum
Graciliala johnsoni
Hercorphycus bicarinata
Liopeplum leioderma
Morea cancellaria
Nonacteonina sp.
Pterocerella pointsettiformis
Pyrifusus sp.

CLASS PELECYPODA

Exogyra ponderosa
Inoceramus (Endocostea) balticus
Pholadomya lincemani
Inoceramus azerbaidjanensis
Inoceramus aff. Barabini
Pinna laqueata?

Table 3. Macrofossil Faunal List

AGE: Upper Middle Campanian Stage/Gulfian Series/Upper Cretaceous
FORMATION: Pecan Gap Chalk/lower Taylor Group
LOCATION: North Sulphur River, northeast Texas

PHYLUM ANNELIDA
Hamulus squasmosus
Hamulus onyx
Proliserpula (Serpula) sp.

PHYLUM ARTHROPODA
CLASS CIRRIPEDIA
ORDER THORACIA
Acroscalpellum sp.

PHYLUM PORIFERA
CLASS CALCAREA (CALCISPONGIAE)
Cliona cretacia

PHYLUM COELENTERATA
CLASS ANTHOZOA
ORDER SCERACTINA
Trochocyathus sp. (solitary form)
Cladophyllia sp. Or *Caulastrea sp* (phaceloid colonial form)

PHYLUM BRYOZOA
CLASS GYMNOLAEMATA
ORDER CHEILOSTOMATA
Coscinopluera (Eschara) sp.?

PHYLUM ECHINODERMATA
CLASS ECHINOIDEA
ORDER SPATANGOIDA
Echinocorys texanus
Hemiaster aff. *lacunosus*
Hemiaster sp.

CLASS CRINOIDEA
SUBCLASS ARTICULATA
FAMILY BOURGUETICRINIDAE or BATHYCRINIDAE
(stem and root fragments only)

PHYLUM MOLLUSCA

CLASS CEPHALOPODA

ORDER AMMONOIDEA

Baculites taylorensis
Baculites mclearni
Hoplitoplacenticeras (Metaplacenticeras) bowersi
Lewyites taylorensis
Trachyscaphites spiniger porchi

ORDER NAUTILOIDEA

Eutrephoceras dekayi (=Nautilus sp. – Stephenson,1918*)*

CLASS GASTROPODA

Acteon sp.
Anchura substriata
Gastrochaena sp.
Gyrodes petrosa
Martesia sp.
Pugnellus densatus
Ringicula pulchella
Schizobasis sp.
Turritella trilira
Volutoderma sp.
Xenophora sp.
Acteon texana
Anchura sp.
Gyrodes major
Liopeplum leioderma
Pluerotomaria sp.
Pyropsis sp.
Sargana sp.
Turritella macnairyensis
Urceolabrum mantachieensis
Xenophora leprosa

CLASS PELECYPODA

Anomia argentaria
Clavagella armata
Cuspidaria aff. *jerseyensis*
Cyprimeria alta
Exogyra ponderosa
Gryphea vesicularis
Inoceramus aff. Barabini
Liospitha protexta
Modiolus sp.
Pecten quinquecostatus
Pteria sp.
Tenea sp.
Cardium sp.
Crassatellites sp.
Cucullaea sp.
Exogyra costata
Exogyra ponderosa erracostata
Inoceramus (Endocostea) balticus
Lima sp.
Liospitha sp.
Pholadomya lincemani
Pecten sp.
Radiolites austinensis
Trigonia sp.

PHYLUM CHORDATA

CLASS CHONDRICHTHYES

SHARKS

Scapanorhynus texanus (=Lamna sp. – Stephenson,1918*)*
Squalicorax kaupi or S.prisodontus (=Corax falcatus – Stephenson,1918*)*

References

Akers, R.E. and Akers, T.J. 1987. Texas Cretaceous Echinoids. Paleontology Section, Houston Gem and Mineral Society, 3:144 p.

Akers, R.E. and Akers, T.J. 1997. Texas Cretaceous Gastropods. Paleontology Section, Houston Gem and Mineral Society, 6:340 p.

Case, G.R. 1982. A Pictorial Guide to Fossils. Van Nostrand Reinhold Company. 516 p.

Cobban, W.A. and Kennedy, W.J. 1992. Campanian ammonites from the Upper Cretaceous Gober Chalk of Lamar County,Texas. Journal of Paleontology, 66(3):440-454.

Cobban, W.A. 1992. Campanian *Trachyscaphites spiniger* ammonite fauna in Northeast Texas. Paleontology, 35(1):63-93.

Cobban, W.A. and Kennedy, W.P. 1993. Middle Campanian ammonites and inoceramids from the Wolfe City Sand in northeastern Texas. Journal of Paleontology, 67(1):71-82.

Dane, C.H., 1929, Upper Cretaceous formations of southwestern Arkansas. Arkansas Geological Survey, 1:215 p.

Emerson, B.L., Emerson, J.H., Akers, R.E., and Akers, T.J. 1994. Texas Cretaceous Ammonites and Nautiloids. Paleontology Section, Houston Gem and Mineral Society, 5:440 p.

Finsley, C. 1989. A field guide to Fossils of Texas. Texas Monthly Press, 190 p.

Gayet, M., Sempere, T., Cappetta, H., Jaillard, E., and Levy, A. 1993. The presence of marine fossils from the late Cretaceous of the central Andes and it's paleogeographic consequences. Palaeogeography, Palaeoclimatology, Palaeoecology, 102:283-319.

Hill, R.T. 1894. Geology of parts of Texas, Indian Territory, and Arkansas adjacent to the Red River. Geologic Society of America Bulletin, 5:297-338.

Larson, N.L., Jorgensen, S.D., Farrar, R.A. and Larson, P.L. 1997. Ammonites and the other cephalopods of the Pierre Seaway. Geoscience Press,Inc., 148 p.

Moore, R.C. (ed). 1969. Treatise on Invertebrate Paleontology (N) – Mollusca 6(2). University of Kansas Press: 751-765.

Moore, R.C. (ed). 1966. Treatise on Invertebrate Paleontology (U) – Echinodermata 3(1). University of Kansas Press: 257-265.

Murray, J.W. (ed). 1985. Atlas of Invertebrate Macrofossils. Halsted Press. 242 p.

Schafer,W. 1972. Ecology and Paleoecology of Marine Environments. University of Chicago Press. 568 p.

Shiner, H.W. and Shrock, R.R. 1944. Index Fossils of North America. John Wiley & Sons, Inc. 838 p.

Stephenson,L.W. 1918. A contribution to the geology of northeastern Texas and southeastern Oklahoma. United States Geological Survey Professional Paper, 120-H:129-163.

Chapter 3 Cretaceous Cephalopods

The large turnout for the Symposium on North Sulphur River (NSR) showed that there is a great interest in all fossil groups. Cephalopods, mainly ammonites and nautiloids, from this area have been well documented. Unfortunately, these publications are scattered throughout the geologic literature. The goal of this chapter is to bring this information together in a single place, to compare published occurrences with the cephalopods brought to the NSR symposium and to provide a simplified identification guide to the ammonites.

AMMONITES

All published NSR ammonite occurrences and synonymies are summarized in Fig. 1 and Appendix 1, respectively. The numbers in the columns of the Figure 1 are tied to the publications at the bottom of the figure. Fig. 1 focuses on the faunas from the rock units which are exposed in the NSR, the Ozan Formation and the Wolfe City Sand. Two other types of faunas are in the figure. It includes the local faunas above and below the Ozan and Wolfe City. These faunas suggest which ammonites may be found in the NSR because of new occurrences or transport from upstream. Text Fig. 1 also has faunas of similar age from outside northeast Texas. These faunas suggest which ammonites may be found in the NSR because of new occurrences.

It is not necessary to repeat all the details from text Fig. 1 and Appendix 1 but a brief summary of the publications is of value. A number of workers have studied fossils from northeast Texas. The first publication on ammonites directly from the NSR was by Young (1963) who did the monographic work on Upper Cretaceous ammonites from Texas. In the collections he studied, Young only reported two of the 13 species currently reported from the NSR. One is clearly from the "redbed" of the Ozan Formation and the other may be from the Gober Chalk. Young reported an additional seven species from formations above and below the Ozan Formation in nearby quarries, railroad cuts, and road cuts. Echols (1984) listed a few ammonites from the same "redbed" but never illustrated them. Finsley (1989) figured and illustrated two species of the ammonites from the NSR area.

Publications, which detail the faunas in the NSR area, are based primarily on the collection of J. P. Conlin, a Dallas collector. Conlin's Ozan Formation ammonites in the United States Geological Survey (USGS) collection were described and illustrated by Cobban and Kennedy (1992a). They also worked on Conlin's ammonites from the concretions of the Wolfe City Sand (Cobban and Kennedy, 1993). These fossils were from a local railroad cut, but can also be found in the North Sulfur River.

A nearly complete listing of NSR ammonites is included in an excellent review of Texas ammonites (Emerson, Emerson, Akers, and Akers, 1994). Unfortunately, they cover only about half of the ammonites in Fig. 1. Five articles on ammonites from Texas and other pertinent US localities were published since 1994.

Although not dealing with the formations exposed in the NSR, there are several publications that list fossils that may yet be found. Cobban and Kennedy (1992b and 1994) documented ammonites from formations above and below the Ozan and Wolfe City. Conlin's collection of the Roxton Member of the Gober Chalk (Cobban and Kennedy, 1992b) added six species to the fauna described by Young (1963). In addition to Conlin's collection, specimens from other collections expanded our knowledge of the Pecan Gap Chalk fauna(Cobban and Kennedy, 1994).

Figure 1. NSR ammonites from the literature

Ammonites SYMBOLS: (1) = REFERENCE* AND <-> = PROJECTED RANGE	Formations Roxton (6)	Ozan (5)	Wolf City (7)	Pecan Gap (9)
Anapachydiscus sp. juv.	6	<->	<->	9
Baculites sp. group aquilaensis Reeside		3,5,10,11		
Baculites haresi Reeside	6,8,10			
Baculites mclearni Landes			7,10	10?
Baculites taylorensis Adkins			11	1,4,9,11
Chesapeakella nodatum Kennedy & Cobban	8			
Cryptotexanites paedomorphicus Kennedy & Cobban	8			
Eubostrychoceras reevesi (Young)				2,9
Eubostrychoceras sp.	6	11		
Eupacydiscus grossouvrei (Kossmat)		5		
Glyptoxoceras sp.	8	5,11		
Hoplitoplacenticeras minor Cobban & Kennedy			2,7,10	
Hoplitoplacenticeras (H.) sp. aff. H. (H.) plasticum Paulcke	6,10		11-?	11-?
Lewyites taylorensis (Adkins)				9
Menabites (Berrella) sp.	8			
Menabites (Delawarella) delawarensis (Morton)	2,6,8	2,3,5,10,11		
Menabites (Delawarella) danei (Young)	2,6,10	5,10,11		
Menabites (Delawarella) vanuxemi (Morton)	2,6-aff,8,10	5,10,11	<->	11
Pacydiscus (Pacydiscus) sp. aff P. haldemsis (Schluter)				9
Pacydiscus (Pacydiscus) paulsoni (Young)	2,6 cf,10	5,10,11		
Pacydiscus (Pacydiscus) travisi (Adkins)			11-cf	1,9,11-cf
Pacydiscus (Pacydiscus) sp.	8			
Parapuzosia terryi Young	2,10			
Patagiosites sp.		10?	7,10	
Placenticeras placenta (DeKay)	6,8,10	11	<->	<->
Placenticeras syrtale (Morton)	2,8,10			
Placenticeras sp.		5	7	
Psuedoschloenbachia cf. P. chispaensis Adkins	8			
Scaphites sp. group hippocrepis (DeKay)		5		
Scaphites (Scaphites) hippocrepis form III (Cobban)	6,8,10			
Submortoniceras punctatum Collignon	8			
Submortoniceras vandaliaense Young	6,10			
Submortoniceras uddeni Young	8			
Submortoniceras sp.	6			
Texanites sp	8			
Trachyscaphites densicostatus Cobban and Kennedy		5,10		
Trachyscaphites spiniger porchi (Adkins)			7,10,11	1,9,10,11
Trachyscaphites spiniger spiniger (Schluter)		3,4,5,11		10?

* References

1 central and west Texas [Adkins, 1929]
2 north Texas [Young, 1963]
3 middle Ozan Fm. [Echols, 1984]
4 north Texas [Finsley, 1989]
5 Ozan Fm. 46m above Gobber Chalk [Cobban and Kennedy, 1992a]
6 Roxton Mem. of Gobber Chalk [Cobban and Kennedy,1992b)
7 Wolfe Cify Sand [Cobban and Kennedy, 1993]
8 Merchantville Fm., NJ [Kennedy and Cobban, 1993]
9 Pecan Gap Chalk [Cobban and Kennedy, 1994]
10 Texas [Emerson, Emerson, Akers, & Akers, 1994]
11 Mississippi & Alabama [Kenney, Cobban, & Landman, 1997]

RWMorin 09-09-99

There are many publications on ammonites which are the same age as those from the NSR. Adkins (1929) and Young (1963) documented ammonites from other parts of Texas. From outside Texas, there are several important faunas of similar ages. The Merchantville Formation of New Jersey, Maryland and Delaware has a fauna similar to the Roxton Member of the Gober Chalk (Kennedy and Cobban 1993). In Mississippi and Alabama, Kennedy, Cobban, and Landman (1997) report faunas which are similar to those of the Ozan Formation, Wolfe City Sand, and Pecan Gap Chalk.

Fig. 2 shows the ammonites brought to the NSR symposium. The columns record ammomite occurrences in several ways. Species found in both the same formation based on the literature and the symposium are marked with an "x". Species not seen at the symposium are indicated by a "?". Species new to the NSR or with extended ranges are marked accordingly. Surprisingly, no ammontes older than the Ozan Formation were found in the NSR. The two Roxton ammonites brought to the symposium were collected in outcrops near the NSR. Ozan ammonites which were collected in the outcrop area of the Wolfe City Sand could be identified by their color and/or matrix. A total of seven Ozan ammonites were identified. The Wolfe City Sand provided one species. Although downstream of the area being studied, two species from the Pecan Gap Chalk were brought to the symposium.

NAUTILOIDS

Nautiloids of the NSR have received less attention than the ammonites. Finsley (1989) listed the genera present in the area. The two species from the Ozan Formation were described by Hanger and Echols (1998). They are *Eutrephoceras dekayi* (Morton) and *Cimomia ozani* Hanger and Echols. *C. ozani* has a slightly sinuous suture while that of E. dekayi has a relatively straight suture. Synonymies are given in Appendix 2.

Only *Eutrephoceras dekayi* (Morton) was brought to the NSR symposium.

Figure 2. NSR ammonites at symposium				
Ammonites	**Formations**			
	Roxton	Ozan	Wolf City	Pecan Gap
Anapachydiscus sp. juv.	?			?
Baculites sp. group aquilaensis Reeside		X		
Baculites haresi Reeside	?			
Baculites mclearni Landes			?	
Baculites taylorensis Adkins			?	?
Chesapeakella nodatum Kennedy & Cobban	?			
Cryptotexanites paedomorphicus Kennedy & Cobban	?			
Eubostrychoceras reevesi (Young)				?
Eubostrychoceras sp.	?	?		
Eupacydiscus grossouvrei (Kossmat)		?		
Glyptoxoceras sp.	?	X		
Hoplitoplacenticeras minor Cobban & Kennedy			?	
Hoplitoplacenticeras (H.) sp. aff. H. (H.) plasticum Paulcke	?		?	?
Lewyites taylorensis (Adkins)				NEW
Menabites (Delawarella) delawarensis (Morton)	X	X		
Menabites (Delawarella) danei (Young)		X		
Menabites (Delawarella) vanuxemi (Morton)	?	?	?	?
Pacydiscus (Pacydiscus) sp. aff P. haldemsis (Schluter)				?
Pacydiscus (Pacydiscus) paulsoni (Young)	X	X		
Para (?) pachydiscus terryi Young	1			
Pacydiscus (Pacydiscus) travisi (Adkins)			?	X
Pacydiscus (Pacydiscus) sp.	?			
Patagiosites sp.		E2A2	?	
Placenticeras placenta (DeKay)	X	X		
Placenticeras syrtale (Morton)	?	EXTEND		
Placenticeras sp.		X	?	
Psuedoschloenbachia cf. P. chispaensis Adkins	?			
Scaphites sp. group hippocrepis (DeKay)		?		
Scaphites (Scaphites) hippocrepis form III (Cobban)	?			
Submortoniceras punctatum Collignon	?			
Submortoniceras vandaliaense Young	?			
Submortoniceras uddeni Young	?			
Submortoniceras sp.	?			
Texanites sp	?			
Trachyscaphites densicostatus Cobban and Kennedy		?		
Trachyscaphites spiniger porchi (Adkins)			X	?
Trachyscaphites spiniger spiniger (Schluter)		X		
Symbols				
X = reported occurrence duplicated at NSR symposium				
? = reported occurrence NOT duplicated at NSR symposium				
NEW = new occurrence to northeast Texas				
EXTEND = extension of range	RWMorin 09-09-99			

References

Adkins, W. S., 1929, Some Upper Cretaceous Taylor ammonites from Texas. University of Texas Bulletin 2901, 203-222.

Echols, J., 1984, Ammonite zonation in condensed zone, middle Ozan Formation (Taylor Group, Upper Cretaceous) in Northeast Texas. AAPG Bulletin, 68 (4), 473.

Cobban, W. A., and Kennedy, W. J., 1992a, Campanian Trachyscaphites spiniger ammonite fauna in Northeast Texas. Palaeontology, 35 (1), 63-93.

Cobban, W. A., and Kennedy, W. J., 1992b, Campanian ammonites from the Upper Cretaceous Gober Chalk of Lamar County, Texas. Journal of Paleontology, 66 (3), 440-454.

Cobban, W. A., and Kennedy, W. J., 1993, Middle Campanian ammonites and inoceramids from the Wolfe City Sand in northeastern Texas. Journal of Paleontology, 67 (1), 71-82.

Cobban, W. A., and Kennedy, W. J., 1994, Middle Campanian (Upper Cretaceous) ammonites from the Pecan Gap Chalk of central and northeastern Texas. William J. (editor), Shorter contributions to paleontology and stratigraphy 1993, U. S. Geological Survey Bulletin, B 2073, D1-D9.

Emerson, B. L., Emerson, J. H., Akers, R. E., and Akers, T. J., 1994, Texas Cretaceous ammonites and nautiloids. Paleontology Section, Houston Gem and Mineral Society, Texas Paleontology Series 5, 439 p.

Finsley, C., 1989, A field guide to fossils of Texas. Texas Monthly Press, 188 p.

Hanger, R. A., and Echols, J., 1998, Nautiloids from the Cretaceous (Campanian) Ozan Formation, Taylor Group, Fannin County, Texas. The Texas Journal of Science, 50 (3), 195-198.

Kennedy, W. J., and Cobban, W. A., 1993, Lower Campanian (Upper Cretaceous) ammonites from the Merchantville Formation of New Jersey, Maryland, and Delaware. Journal of Paleontology, 67 (5), 828-849.

Kennedy, W. J., Cobban, W. A., and Landman, N. H., 1997, Campanian ammonites from the Tombigbee Sand Member of the Eutaw Formation, the Mooreville Formation, and the basal part of the Demopolis Formation in Mississippi and Alabama. American Museum Novitates, 3201, 44 p.

Kennedy, W. J., Cobban, W. A., Landman, N. H., and Johnson, R. O., New ammonoid records from the Merchantville Formation (Upper Cretaceous) of Maryland and New Jersey. American Museum Novitates, 3193, 17 p. (not found)

Young, K, 1958, Age of the Austin Chalk (Cretaceous). Geol. Soc. America, 69 (2), 1669.

Young, K., 1963, Upper Cretaceous ammonites from the Gulf Coast of the United States. Texas Univ. Pub. 6304, 373 p.

APPENDIX 1: AMMONITE SYNONOMY (ALPHABETIC)

Acanthoscaphites sp. cfr. A. spiniger porchi (Adkins); Young (1963)
 see Trachyscaphites spiniger porchi (Adkins)
Anapachydiscus sp. juv.
 1992b Anapachydiscus sp. juv.; Cobban and Kennedy, p. 443, Roxton
 1994 Anapachydiscus sp. juv.; Cobban and Kennedy, p. D4, Pecan Gap
Baculites sp. group of aquilaensis Reeside
 1984 Baculites aquilaensis Reeside; Echols, p. 473, middle Ozan
 1992a Baculites sp. group of aquilaensis Reeside; Cobban and Kennedy, p. 82 , Ozan
 1994 Baculites sp. cfr. B. aquilaensis Reeside; Emerson, Emerson, Akers, and Akers, p. 321, Ozan
 1997 Baculites sp. group of aquilaensis Reeside ; Kennedy, Cobban, and Landman, p. 36, Arcola
Baculites haresi Reeside
 1992b Baculites haresi Reeside; Cobban and Kennedy, p. 449, Roxton
 1993 Baculites haresi Reeside; Kennedy and Cobban, p. 844, Merchantville
 1994 Baculites haresi Reeside; Emerson, Emerson, Akers, and Akers, p. 322, Gober
Baculites mclearni Landes
 1984 Baculites mclearni Landes; Echols, p. 473, identification questioned, left off spreadsheet
 1984 Baculites obtusus Meek; Echols, p. 473, identification questioned , left off spreadsheet
 1993 Baculites mclearni Landes; Cobban and Kennedy, p. 75, Wolfe City
 1994 Baculites mclearni Landes; Emerson, Emerson, Akers, and Akers, p. 322, Wolfe City and Pecan Gap (not reported by Cobban and Kennedy, 1994)
Baculites obtusus Meek, Echols (1984)
 see Baculites mclearni Landes
Baculites taylorensis Adkins
 1929 Baculites taylorensis Adkins, p. 204. Pecan Gap
 1989 Baculites taylorensis Adkins: Finsley, p. 83, Pecan Gap
 1994 Baculites taylorensis Adkins: Cobban and Kennedy, p. D6, Pecan Gap
 1994 Baculites taylorensis Adkins; Emerson, Emerson, Akers, and Akers, p. 323, Taylor
 1997 Baculites taylorensis Adkins; Kennedy, Cobban, and Landman, p. 36, Demopolis
Chesapeakella nodatum Kennedy & Cobban
 1993 Chesapeakella nodatum Kennedy & Cobban; Kennedy and Cobban , p. 844, Merchantville
Cirroceras reevesi Young
 see Eubostrychoceras reevesi (Young)
Cryptotexanites paedomorphicus Kennedy & Cobban
 1993 Cryptotexanites paedomorphicus Kennedy & Cobban; Kennedy and Cobban, p. 844, Merchantville
Delawarella delawarensis (Morton)
 see Menabites (Delawarella) delawarensis (Morton)
Delawarella danei (Young)
 see Menabites (Delawarella) danei (Young)
 1994 Didymoceras reevesi (Young); Emerson, Emerson, Akers, and Akers, p. 312, Anacacho
Eubostrychoceras sp.
 1992b Eubostrychoceras sp.; Cobban and Kennedy, p. 449, Roxton
 1997 Eubostrychoceras sp; Kennedy, Cobban, and Landman, p. 35, Arcola
Eupacydiscus grossouvrei (Kossmat)
 1992a Eupacydiscus grossouvrei (Kossmat); Cobban and Kennedy, p. 70, Ozan
 1994 Eupacydiscus grossouvrei (Kossmat); Emerson, Emerson, Akers, and Akers, p. 82, Ozan

Glyptoxoceras sp.

1992a Glyptoxoceras sp.; Cobban and Kennedy, p. 80, Ozan

1993 Glyptoxoceras sp.; Kennedy and Cobban, p. 844, Merchantville

1997 Glyptoxoceras sp; Kennedy, Cobban, and Landman, p. 35, Arcola

Hoplitoplacenticeras marroti (Coquand)

see Hoplitoplacenticeras minor Cobban & Kennedy

Hoplitoplacenticeras minor Cobban & Kennedy;

1963 Hoplitoplacenticeras marroti (Coquand) Young, p. 63 (parts), Wolfe City

1993 Hoplitoplacenticeras minor Cobban & Kennedy; Cobban and Kennedy, p. 73, Wolfe City

1994 Hoplitoplacenticeras minor Cobban & Kennedy; Emerson, Emerson, Akers, and Akers, p. 103, Wolfe City

1994 Hoplitoplacenticeras marroti (Coquand); Emerson, Emerson, Akers, and Akers, p. 103, Wolfe City

Hoplitoplacenticeras (H.) sp. aff. H. (H.) plasticum Paulcke

1992b Hoplitoplacenticeras (H.) sp. aff. H. (H.) plasticum Paulcke; Cobban and Kennedy, p. 443, Roxton

1994 Hoplitoplacenticeras (H.) sp. aff. H. (H.) plasticum Paulcke; Emerson, Emerson, Akers, and Akers, p. 103, Gober

1997 Hoplitoplacenticeras (H.) sp.; Kennedy, Cobban, and Landman, p. 15, Demopolis

Lewyites taylorensis (Adkins)

1929 Hamites (?) taylorensis Adkins; p. 209, Taylor

1994 Lewyites taylorensis (Adkins); Emerson, Emerson, Akers, and Akers, p. 317, Taylor

1994 Lewyites taylorensis (Adkins); Cobban and Kennedy, p. D5, Pecan Gap (central Texas)

Menabites (Berrella) sp.

1993 Menabites (Berrella) sp.; Kennedy and Cobban, p. 841, Merchantville

1994 Menabites (Berrella) sp.; Emerson, Emerson, Akers, and Akers, p. 246, Texas

Menabites (Delawarella) delawarensis (Morton)

1963 Delawarella delawarensis (Morton); Young, p. 111, Gober and Ozan

1984 Delawarella delawarensis (Morton); Echols, p. 473, middle Ozan

1992a Menabites (Delawarella) delawarensis (Morton); Cobban and Kennedy, p.73, Ozan

1992b Menabites (Delawarella) delawarensis (Morton); Cobban and Kennedy, p. 445, Roxton

1993 Menabites (Delawarella) delawarensis (Morton); Kennedy and Cobban , p. 836, Merchantville

1994 Menabites (Delawarella) delawarensis (Morton); Emerson, Emerson, Akers, and Akers, p. 244, Ozan

1997 Menabites (Delawarella) delawarensis (Morton); Kennedy, Cobban, and Landman, p. 23, Arcola

Menabites (Delawarella) danei (Young)

1963 Delawarella) danei Young, p. 114, Gober Chalk

1992a Menabites (Delawarella) danei (Young); Cobban and Kennedy, p. 74, Ozan

1992b Menabites (Delawarella) danei (Young); Cobban and Kennedy, p. 448, Roxton

1994 Menabites (Delawarella) danei (Young); Emerson, Emerson, Akers, and Akers, p. 244, Gober and Ozan

1994 Menabites (Delawarella) vanuxemi (Morton); Emerson, Emerson, Akers, and Akers, p. 244, Gober and Ozan

Menabites (Delawarella) sp. aff. M. (D.) vanuxemi (Morton)

1992b Menabites (Delawarella) sp. aff. M. (D.) vanuxemi (Morton); Cobban and Kennedy, p. 448, Roxton

Pacydiscus (Pacydiscus) sp. aff P. haldemsis (Schluter)

1994 Pacydiscus (Pacydiscus) sp. aff P. haldemsis (Schluter): Cobban and Kennedy, p. D3, Pecan Gap

Pacydiscus (Pacydiscus) paulsoni (Young)

1963 Parapacydiscus paulsoni Young; Young, p. 53, Gober Chalk

1992a Pacydiscus (Pacydiscus) paulsoni (Young); Cobban and Kennedy, p. 69, Ozan

1992b Pacydiscus (Pacydiscus) sp. cf. paulsoni (Young); Cobban and Kennedy, p. 441, Roxton

1994 Pacydiscus (Pacydiscus) paulsoni (Young); Emerson, Emerson, Akers, and Akers, p.78, 79, Gober and Ozan

1997 Pacydiscus (Pacydiscus) paulsoni (Young); Kennedy, Cobban, and Landman, p. 9, Arcola

Pacydiscus (Pacydiscus) travisi (Adkins)

1929 Parapachcydiscus travisi Adkins, p. 207, Pecan Gap

1994 Pacydiscus (Pacydiscus) travisi (Adkins); Cobban and Kennedy, p. D3, Pecan Gap

1994 Pacydiscus (Pacydiscus) travisi (Adkins); Emerson, Emerson, Akers, and Akers, p. 78, Taylor

1997 Pacydiscus (Pacydiscus) travisi (Adkins); Kennedy, Cobban, and Landman, p. 8, Demopolis

Pacydiscus (Pacydiscus) sp.

1993 Pacydiscus (Pacydiscus) sp.; Kennedy and Cobban, p. 831, Merchantville

Parapachcydiscus travisi Adkins

see Pacydiscus (Pacydiscus) travisi (Adkins)

Parapuzosia terryi Young

1963 Parapuzosia terryi Young, p. 53, Gober

1994 Parapuzosia terryi Young; Emerson, Emerson, Akers, and Akers, p. 62, Gober

Patagiosites sp.

1993 Patagiosites sp.; Cobban and Kennedy, p. 72, Wolfe City

1994 Patagiosites sp.; Emerson, Emerson, Akers, and Akers, p. 70, 74, Ozan? (Wolfe City reported by Cobban and Kennedy, 1993)

Placenticeras placenta (DeKay)

1989 Placenticeras whitefieldi Hyatt; Finsley, p. 84, Taylor Group

1992b Placenticeras placenta (DeKay); Cobban and Kennedy, p. 443, Roxton

1993 Placenticeras placenta (DeKay); Kennedy and Cobban , p. 834, Merchantville

1994 Placenticeras placenta (DeKay); Emerson, Emerson, Akers, and Akers, p. 97, Gober

1997 Placenticeras placenta (DeKay); Kennedy, Cobban, and Landman, p. 10, Arcola

Placenticeras syrtale (Morton)

1963 Stantonoceras sancarlosense; Young, p. 63, Gober Chalk equivalent (Oklahoma?)

1993 Placenticeras syrtale (Morton); Kennedy and Cobban, p. 835, Merchantville

1994 Placenticeras syrtale (Morton); Emerson, Emerson, Akers, and Akers, p. 97, 98, Gober

Placenticeras whitefieldi Hyatt, Finsley (1989)

see Placenticeras placenta (DeKay)

Placenticeras sp.

1992a Placenticeras sp.; Cobban and Kennedy, p. 73, Ozan

1993 Placenticeras sp; Cobban and Kennedy, p. 72, Wolfe City

Scaphites aricki Adkins

see Trachyscaphites spiniger porchi (Adkins)

Scaphites sp. group hippocrepis (DeKay)

1989 Scaphites hippocrepis (DeKay); Finsley, p 83, Upper Cretaceous

1992a Scaphites sp. group hippocrepis (DeKay); Cobban and Kennedy, p. 84, Ozan

Scaphites (Scaphites) hippocrepis form III (Cobban)

Psuedoschloenbachia cf. P. chispaensis Adkins

1993 Psuedoschloenbachia cf. P. chispaensis Adkins; Kennedy and Cobban , p. 832, Merchantvill

1992b Scaphites (Scaphites) hippocrepis form III (Cobban); Cobban and Kennedy, p. 451, Roxton

1993 Scaphites (Scaphites) hippocrepis form III (Cobban); Kennedy and Cobban , p. 845, Merchantville

1994 Scaphites (Scaphites) hippocrepis (Dekay); Emerson, Emerson, Akers, and Akers, p. 329, Gober

Scaphites porchi Adkins

see Trachyscaphites spiniger porchi (Adkins)

Stantonoceras sancarlosense (Hyatt), Young (1963)

see Placenticeras syrtale (Morton)

Submortoniceras punctatum Collignon

1993 Submortoniceras punctatum Collignon; Kennedy and Cobban , p. 841, Merchantville

Submortoniceras vandaliaense Young

1963 Submortoniceras vandaliaense Young, p. 102, Bloosom Sand

1992b Submortoniceras vandaliaense Young; Cobban and Kennedy, p. 448, Roxton

1994 Submortoniceras vandaliaense Young; Emerson, Emerson, Akers, and Akers, p. 236, 238, Blossom

Submortoniceras uddeni Young

1993 Submortoniceras uddeni Young; Kennedy and Cobban, p. 842, Merchantville

1994 Submortoniceras uddeni Young; Emerson, Emerson, Akers, and Akers, p. 238, Austin

Submortoniceras vanuxemi (Morton)

see Menabites (Delawarella) vanuxemi (Morton)

Submortoniceras sp.

1992b Submortoniceras sp.; Cobban and Kennedy, p. 449, Roxton

Texanites sp.

1993 Texanites sp.; Kennedy and Cobban, p. 836, Merchantville

Trachyscaphites densicostatus Cobban and Kennedy

1992a Trachyscaphites densicostatus Cobban and Kennedy; Cobban and Kennedy, p. 88, Ozan

1994 Trachyscaphites densicostatus Cobban and Kennedy; Emerson, Emerson, Akers, and Akers, p. 337, Ozan

Trachyscaphites spiniger ariki (Adkins); Finsley (1989)

see Trachyscaphites spiniger spiniger (Schluter)

Trachyscaphites spiniger porchi (Adkins)

1929 Scaphites porchi Adkins, p. 205, Taylor and Pecan Gap (central Texas)

1929 Scaphites aricki Adkins, p. 206, Taylor and Pecan Gap (central Texas)

1963 Acanthoscaphites sp. cfr. A. spiniger porchi (Adkins; Young, p. 49, Anacacho

1993 Trachyscaphites spiniger porchi (Adkins); Cobban and Kennedy, p. 77, Wolfe City

1994 Trachyscaphites spiniger porchi (Adkins); Cobban and Kennedy, p. D7, Pecan Gap

1994 Trachyscaphites spiniger porchi (Adkins); Emerson, Emerson, Akers, and Akers, p. 337, Pecan Gap and Wolfe City

1997 Trachyscaphites spiniger porchi (Adkins); Kennedy, Cobban, and Landman, p. 40, Demopolis

Trachyscaphites spiniger spiniger (Schluter)

1984 Trachyscaphites spiniger porchi (Adkins); Echols, p. 473, identification questioned, middle Ozan

1989 Trachyscaphites spiniger ariki (Adkins); Finsley, p. 83, identification questioned, Taylor Group

1992a Trachyscaphites spiniger spiniger (Schluter); Cobban and Kennedy, p. 86, Ozan

Uncertain species:

Submortoniceras aff. condamyi (Collignon)

1958 Submortoniceras aff. condamyi (Collignon); Young, p. 1669, Austin & Ozan

1994 Submortoniceras aff. condamyi (Collignon); Emerson, Emerson, Akers, and Akers, p. 237, Ozan

Hoplitoplacenticeras (H.) sp. aff. Metaplacenticeras boweri Anderson

1949 Hoplitoplacenticeras sp. aff. vari; Adkins in Feray and Plummer, p. 62, Pecan Gap

1963 Hoplitoplacenticeras sp. aff. Metaplacenticeras boweri Anderson; Young, p. 64, Pecan Gap

1994 Hoplitoplacenticeras (H.) sp. aff. Metaplacenticeras boweri Anderson; Emerson, Emerson, Akers, and Akers, p. 101, Pecan Gap

Neancylocers clinensis (Adkins)

1929 Hamites (?) clinensis Adkins, p. ?, Anacacho

1960 Neancylocers clinensis (Adkins); Young, p. ?, Anacacho

APPENDIX 2: NAUTILOID SYNONOMY (ALPHABETIC)

Cimomia ozani Hanger and Echols

1989 Cimomia spp.: Finsley, p 81, Upper Cretaceous

1998 Cimomia ozani Hanger and Echols, p. 196, Ozan

Eutrephoceras dekayi (Morton)

1989 Eutrephoceras spp. ; Finsley, p. 81, Paleocene-Eocene (sic)

1994 Eutrephoceras dekayi (Morton); Emerson, Emerson, Akers, and Akers, p. 45, Navarro

1998 Eutrephoceras dekayi (Morton); Hanger and Echols, p. 197, Ozan

KEY TO NORTH SULFUR RIVER AMMONITES

Numbers in parentheses after a genus are pages for description in Treatise of Cretaceous Ammonites (Wright, C. W., Calloman, J. H., Howarth, M. K., Mollusca, 1996, Cretaceous Ammonoidea, Treatise on invertebrate paleontology, L (4), 362 p.)

Normal coiling in a plane (planispiral) I

no keel (not carinate)... 1

keel (carinate) 2

Partly to totally uncoiled (hetermorphs) II

Normal coiling in a plane (planispiral) I

no keel(not carinate) 1

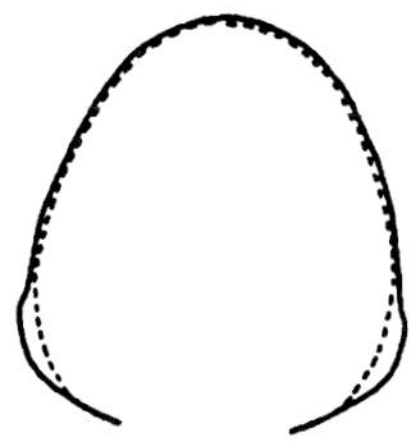

ribs cross venter

constrictions persist ....**Patagiosites** (70)

no constrictions

wide cross section

weak ribs .... **Anapachydiscus** (75)

strong ribs .... **Eupacydiscus** (82)

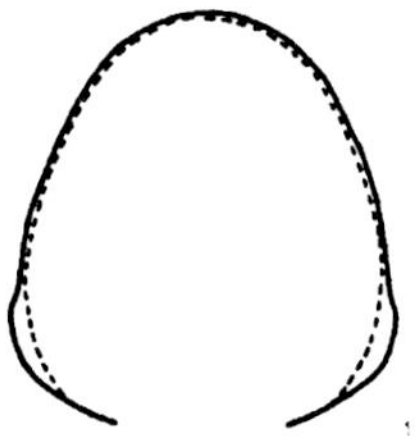

tall cross section ....**Pachydiscus** (78)

P. (P.) paulsoni ... feebly bullate primary ribs
P. (P.) travisi ...... prominent bullae

ribs do not cross venter

Weak to no ornament .... **Placenticeras** (97)

P. placenta .... unornamented,
P. syrtale ...... ornamented, bullae and clavi

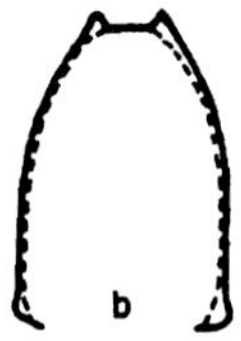

Prominent ribs

elongate tubercles ...... **Hoplitoplacenticeras** (103)

H. minor ....1 or 2 secondary ribs
H. sp. aff. H. plasticum ... few secondary ribs

spinose tubercles .... **Stantonoceras** (~97)

Normal coiling in a plane (planispiral) I

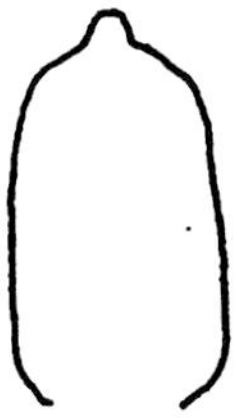

keel (carinate) 2

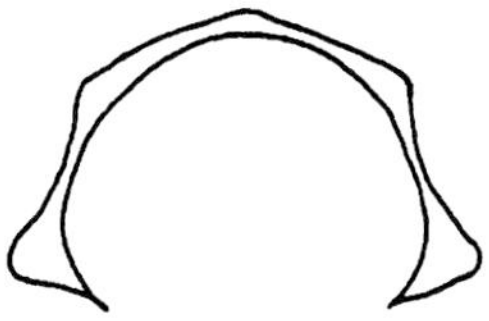

1 to 2 tubercles ....**Psuedoschloenbachia** (86)

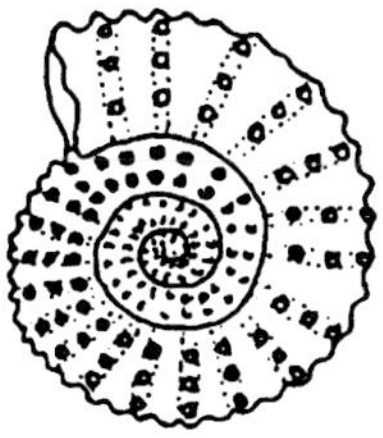

3 to 5 tubercules

dwarf .... **Cryptotexanites** (—)

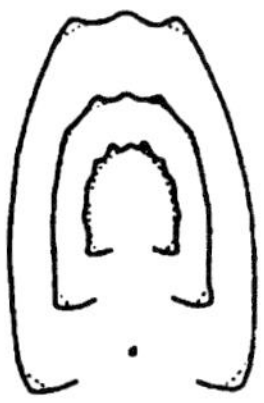

only umbilical tubercles late ... **Submortoniceras** (237,8)

S. punctatum .... strongest ornamentation
S. vandaliaense .. texanitine costae disappear early
S. uddeni .... texanitine clavae persistent

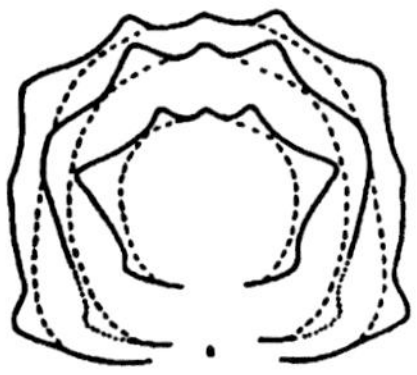

3 tubercles early .... **Menabites (Delawarella)** (244)

M. (D.) danei .... coarse ornament, trituberculate to 150 mm
M. (D.) delawarensis .. moderate ornament
M. (D.) vanuxemi ......crowded ribs, weak effacing tubercles
M. (D.) sp. aff. M. (D.) vanuxemi (Morton) .. juvenile

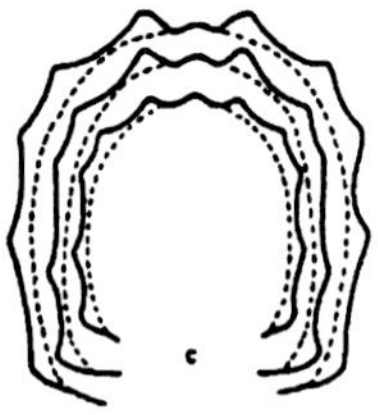

5 tubercles .... **Texanites** (233)

partly to totally uncoiled (hetermorphs) II

coiled in one plane

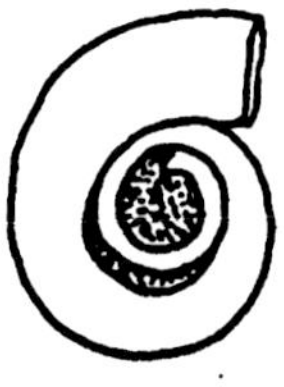

open coil

0 tubercles .... **Glyptoxoceras** (315)
1 tubercle .... **Chesapeakella** (—)
2 tubercles .... **Lewyites** (317)

coiled portion prominent

1-2 tubercles .... **Scaphites** (329)
3-5 tubercles .... **Trachyscaphites** (337)

T. densicostatus .... three rows of tubercles
T. spiniger porchi ... five rows of tubercles, lacks ribs on most of the body chamber
T. spiniger spiniger .. five rows of tubercles, ribs on most of the body chamber

small coil, mainly straight shaft .... **Baculites** (321-3)

B. sp. group aquilaensis Reeside ... rapid taper
B. haresi............... coarse ventral and ventrolateral ribbing only
B. obtusus .... dorsal ribs are sparse and distinct
B. taylorensis ...... strong node-like bullae

not coiled in one plane

initial spiral .... **Eubostrychoceras** (309)

Chapter 4 Cretaceous Vertebrates

One of the main attractions of fossil collecting along the NSR is the abundance of vertebrate material. Most Cretaceous vertebrate remains are isolated mosasaur vertebrae, fish and shark teeth, crocodile scutes, and miscellaneous bone scrap from the gravel bars. Occasionally, more unusual or complete specimens are found. Let's face it! That is what keeps drawing us back to the river time and time again. The potential for finding the big one that didn't get away! The reason most of the vertebrate material consits of isolated specimens is that it is has been previously weathered out of the matrix and transported to its present location on the gravel bar. The diligent collector must search the river and creek banks themselves for signs of bone material still in situ.

Tables 4, 5 and 6 are listings of late Santonian through mid Campanian /Upper Cretaceous vertebrates from North America. Table 4 is marine tetrapods, Table 5 is marine fishes, and Table 6 is marine sharks and rays. All three tables are reformatted from the work by Nicholls and Russell (1990). In their original paper they did not tabulate vertebrate data from the Ozan Formation in Texas because of uncertainties as to the exact age of the material collected. In this guidebook we have included all previously cited references of Cretaceous vertebrates from the Ozan Formation plus any first reported occurrences as a result of the NSR Symposium. The relative age of the Ozan Formation is no longer in doubt. Cobban and Kennedy, in a series of papers on the ammonite fossils of northeast Texas (1992,1993) successfully demonstrated that the Ozan, Wolfe City, and Pecan Gap formations are all mid Campanian in age. This correlates with the upper lower Campanian of earlier literature. Refer to Fig.2 in Chapter 1 as to the relative ages of these six localities. They range in age from late Santonian (oldest) to mid Campanian (youngest).

A cursory review of the three tables indicates that the Ozan Formation of Texas is very probably under evaluated relative to its age equivalent neighbors in the United States. Note in Table 4 and 5 the abundance of marine tetrapods and fish from the Niobrara Chalk to the north in Kansas and the Mooreville Chalk to the east in Alabama. In Table 5, there are 31 reported fish genera from the Niobrara Chalk, 16 from the Mooreville Chalk, and only 9 from the Ozan Formation. Seven new fish genera were recorded from the Ozan; *Pachyrhizodus, Gillicus, Hoploptyrex, Anomoeodus, Saurodon, Stephanodus* and *Protosphyraena*. In addition, unidentified bladed and pycnodont fish teeth are present. The ichthyodectids are a family of medium to large-sized primitive fish that includes *Xiphactinus, Gillicus*, and *Ichthyodectes*. Their vertebrae are all very similar in appearance and are easily misidentified for one another. I can guarantee that Ichthyodectes should be on the list of fish from the Ozan Formation. But this bony fish has yet to be positively identified and so is not listed.

Russell (1993) described and tabulated the marine vertebrate faunas from the Western Interior Sea for the whole Cretaceous Period. The Cretaceous can be subdivided into four broad provincial ages based on marine vertebrate assemblages. In ascending order they are Trinitian, Woodbinian, Niobraran, and Navesinkan. The Trinitian "age" is based on vertebrates from the early Cretaceous Trinity Series in north central Texas. It spans the Aptian to early Cenomanian time interval and lasted approximately 20 million years. The Woodbinian "age" is based on Woodbine vertebrate deposits from north central Texas. It is a transitional stage of short duration (a few million years) during the middle Cenomanian. The Niobraran "age" is characterized by the abundant marine vertebrate assemblages from the Niobrara Chalk of west

Kansas. It lasted from the latest Cenomanian to the mid Campanian Stage. This is a time interval of roughly 15 million years. In Texas this includes the Kamp Ranch and Arcadia Park Members of the Eagle Ford Group, the entire Austin Group, and the lower Taylor Group up to the top of the Ozan Formation. The uppermost "age" based on marine vertebrate assemblages is the Navesinkan. This grouping is based on vertebrate deposits from the Navesink Formation in New Jersey. It lasted until the close of the Cretaceous, a time interval of 15 million years. The Pecan Gap Chalk and younger Cretaceous units in Texas belong to this interval.

The Niobraran "age" was characterized by an abundance of large teleost fish, sea turtles, the dominance of mosasaurs over plesiosaurs and pliosaurs, and the diversity in hesperornithiformes (sea birds). It includes the appearance of such well known fossil fish as *Xiphactinus, Protosphyraena,* and *Apsopolix.* The Western Interior Seaway was an open migration route for subtropical faunas to enter from the south (Texas) and temperate faunas from the north (Canada).

The apparent low diversity of vertebrate faunas in Texas is the combination of numerous factors. It is partially a sampling bias caused by the lack of articulated specimens which makes species determination more difficult. Imagine the inherent difficulty in positively identifying a species of mosasaur based on a single tooth or a turtle based on only a scute. For example, outside of a *Globidens* tooth, most mosasaur teeth basically look alike. Abundant vertebrate material is found in the NSR but not a lot of <u>identifiable</u> vertebrate material is found.

Another potential sampling bias involves the method by which fossils are collected from the NSR. Walking the gravel bars is a good way to spot the big stuff such as mosasaur vertebrae or Pleistocene bison bones. It is not necessarily the best way to collect the smaller material that may be more diagnostic for identification purposes. This is especially true in regards to fish teeth. A majority of the fish teeth are very slender and under one inch in length. They are easily overlooked when walking the gravel bars.

The exception to the apparent paucity of vertebrates from the Ozan Formation is recorded in Table 6. The excellent monograph by DPS members Welton and Farish (1993) recorded 22 genera of sharks, skates and rays from the Campanian of Texas. This is slightly more than the 18 recorded genera from the Mooreville Chalk of Alabama. This is a good indication that a focused effort on any group of vertebrates from the NSR will ultimately lead to a much more comprehensive listing of Cretaceous marine vertebrates than currently available.

During the late Cretaceous, northeast Texas was located offshore from a peninsula of land that separated the Western Interior Seaway from the reentrant Mississippi Embayment to the east (see Fig.6). It lay at the southern terminus of the seaway where it opened into the present-day Gulf of Mexico. It is logical to assume that there was an open exchange of marine faunas from the Tethys Sea into the Western Interior Seaway throughout the later part of the Cretaceous Period. Marine vertebrates today are cosmopolitan in nature as far as their distribution is concerned. For example, the California Gray Whale winters in Baja California off the coast of Mexico but migrates all the way into the arctic Bering Sea during the summer months in search of food. It gives birth to live young and must surface to breath air just like us. Whale ancestors were land-dwelling creatures that migrated back into the ocean during the early part of the Tertiary. One of the reasons they were able to go "back to the sea" is that the marine reptilian predators went extinct at the end of the Cretaceous. The "top of the food chain" were carnivores like the mosasaurs and plesiosaurs.

The killer whale *Orca* is probably a good analog for a mosasaur. It is a large and efficient marine killing machine. Of course, it is a mammal and not a reptile. The origin of mosasaurs dates back into the Cenomanian from Europe. Interestingly enough, the oldest mosasauroid remains in North America come from the Ernst Member of the Boquillas Formation in the Big Bend area of west Texas. These are Middle Turonian in age. Mosasaurs arrived later to the Western Interior Seaway than the plesiosaurs but were ultimately more successful. By the Campanian the mosasaurs were the dominant large carnivore in theCretaceous oceans. The tylosaurs were the largest of the mosasaurs. The six-foot skull that Jim Chaney brought to the NSR Symposium is probably a species of *Tylosaurus* (Plate 14, Fig.1-3). If the skull to body ratio of 8-10 is applied, his specimen would have been 48 to 60 feet long. This would make it one of the largest mosasaurs found anywhere in the world!

Mosasaurs bore their offspring live in the water. Recent discoveries from South Dakota ends the debate if they layed eggs on land or if they gave birth to live young that stayed with them for at least the first year of life.

Mosasaur dentition is the key to their dietary habits. The size and shape of their teeth scream carnivore. Mosasaur bite marks and coprolites indicate that they fed on all types of prey. Ammonites, squids, fish, turtles, plesiosaurs, other mosasaurs and even birds and pteranodons were considered fair game. They in turn were fed upon by scavenging sharks such as *Squalicorax* and *Cretoxyrhina.* Killer whales have been recorded on film using their blunt snouts as battering rams to stun their prey. They sometimes appear to make a game out of hurling sea lions and seals into the air before going in for the kill. It has been postulated that the massive bony snout of the tylosaurs was used for the same purpose.

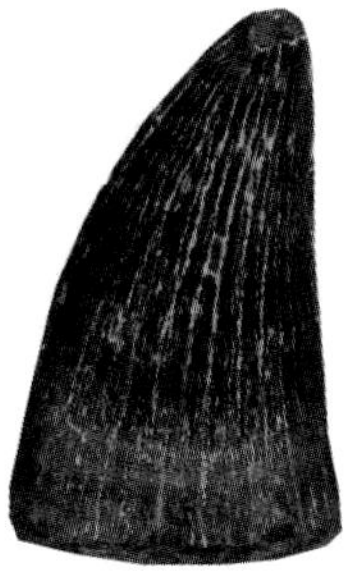

Fig. 1 *Tylosaur* tooth- meat eater [same scale]

Fig. 2 *Globidens* tooth – shell crusher

An all together different lifestyle has been attributed to *Globidens* (see Plate 12, Fig.2). It has globular teeth that are radically different in shape from other mosasaurs. It has been suggested that it was adapted to feeding on molluscs along the sea bottom. Its rounded teeth were probably used for crushing and grinding shells and then swallowing the remains.

Note the abundance of chelonioids (marine turtles) in Table 4 from the Niobrara and the Mooreville Chalks. There are 11 reported species of sea turtles from Alabama and 13 species from Kansas. Six species are found in both deposits. They are *Toxochelys latiremis, Ctenochelys steonoporus, Prionochelys nauta, Chelosphargis advena, Calcarhichelys gemma,* and *Protostega gigas* per Hirayama (1997). All these species have the potential of being discovered from the Ozan Formation in northeast Texas. Once again, it is probably more a matter of not having identifiable material and a lack of local expertise on Cretaceous marine turtles.

The most common fossil remains of marine turtles from the NSR are isolated scutes (Plate 16, Figs. 5 and 6). Chelonioids appeared suddenly in the early Cretaceous approximately 110 MYA in the late Aptian/early Albian. Undescribed remains are known from the Glen Rose Formation in central Texas. There are only seven extant species. Modern chelonioids are primarily restricted to tropical and subtropical waters. They come ashore to lay eggs on sandy beaches at night under the cover of darkness. Modern turtles are passive creatures that feed on a diet of algae, seaweed, and jellyfish. Teeth are lacking or much reduced. *Protostega* and *Archelon* were giant sea turtles with carapaces up to ten feet in diameter.

Two unidentifiable dinosaur bone pieces were brought to the NSR Symposium (Plate 16, Fig.7 and 8). These were identified as dinosaurian by the very large size of the specimens and the density of the outer bone versus the interior bone marrow. Occasionally, a dinosaur carcass would wash out to sea during a flooding event. Longshore currents would eventually carry the corpse away from shore to its watery grave. If conditions were just right the carcass could settle to the bottom before scavenging animals totally dismembered the body and picked it clean. This is the "bloat and float" theory for terrestrial dinosaur remains in a marine deposit. Such was the case when Monte Payton discovered a battery of hadrosaur teeth from the Ozan Formation in Fannin County, Texas. There is a very nice picture of the find on page 51 of Charles Finsley's 1999 book on "Discover Texas Dinosaurs". The specimen now resides at the vertebrate lab at UT Austin.

It is readily apparent from Table 5 that the marine fishes are woefully under represented from the NSR. This is probably due to a lack of readily identifiable specimens. *Enchodus* palatine teeth and isolated *Xiphactinus* teeth are common from the gravel bars. Both of these fish were predators. *Enchodus* has been called the Sabertooth Herring of the Cretaceous Seas because of its distinctive, curved palatine teeth at the front of its upper jaw (see Plate 10, Fig.4-6). *Xiphactinus* (Plate 10, Fig.8) was one of the largest teleost predators in the Western Interior Sea. It reached overall body lengths of six meters and is common in the Smokey Hill Member of the Niobrara Chalk in western Kansas. *Xiphactinus* had a bulldog shaped mouth with large fangs at the front on its premaxillary bones. These small bones were movable and allowed the jaws to extend open very wide. *Xiphactinus* was the ultimate "bigmouth". Evidence of just how much it could swallow can be seen at the recently reopened Sternberg Museum in Hays, Kansas. The museum has a wall mount of a large *Xiphactinus* that has swallowed a six foot *Gillicus* fish whole! Of course, the meal gave it a case of terminal indigestion.

Figs. 3 & 4 Head region from a restored cast of a *Xiphactinus audax* skeleton and the original fossil in matrix from Coniacian age sediments within the Niobrara Chalk from Western Kansas. This is the same species whose isolated teeth and vertebrae are common along the NSR. Note the massive jaws and teeth of varying size.

Protosphyraena (see Plate 11, Fig.5) is an unusual fish that has been called the Swordfish of the Cretaceous Seas. It had a long pointed stout (rostrum) much like the modern day swordfish. It is a characteristic fish of the Niobraran "age" in North America. Skulls with the rostrum attached have been collected by DPS members in the Smokey Hill Member of the Niobrara Chalk in west Kansas.

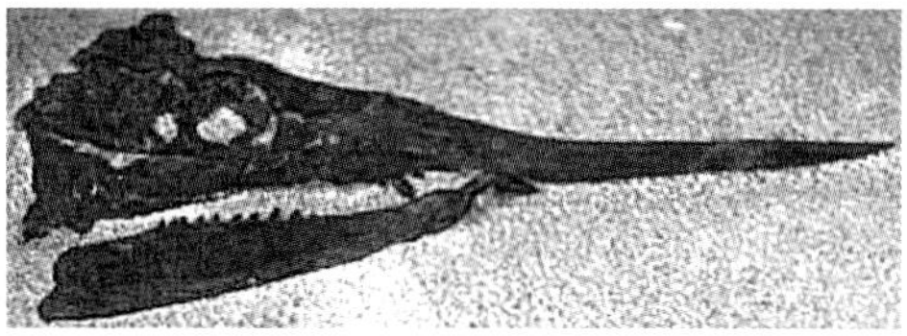

Fig. 5 A complete skull of *Protosphyraena sp.* in right lateral view that clearly shows the elongated rostrum that gives this genus the nickname of "Swordfish of the Cretaceous Seas". From the Niobrara Chalk of Kansas.

Another group of fishes with easily recognized teeth is the pycnodonts (see Plate 10, Fig.3). These were small to medium sized fish that were laterally compressed and deep bodied (see text Fig.6 below). This body shape is designed for slow, adept, deliberate movements through potentially tight places like cavities in coral reefs. Their dentary is characterized of nipping teeth at the front of the mouth and dental pavements composed of rows of small, round to rectangular, peg-like teeth in the back. Pycnodonts nibbled on invertebrates along the sea floor and used their batteries of teeth to pulverize the shell material into edible pieces. They were a highly successful group that reached their peak in the early Cretaceous. Theirfossil remains have been found in both reef and in muddier, softer sediments. It has been proposed that pycnodonts scraped off the algae from corals and other invertebrates. If you have ever scuba dived in the Caribbean, you may have witnessed this same type of feeding behavior by angelfishes, butterflyfishes, and surgeonfishes.

Fig. 6 A complete pycnodont fish with its characteristic body shape and small mouth that is lined with pebble-like teeth. This is *Microdon sp.* from the Jurassic.

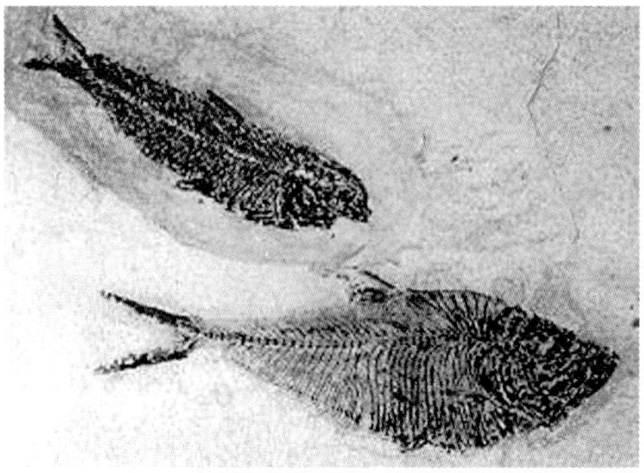

Fig. 7 2 specimens of the herring shaped *Hoplopteryx sp.* from the Cenomanian Stage English Chalk.

Here is an exmple of why it is so tough to identify fish species based on partial fossil remains. Fig. 8 shows the left prearticular from three seperate *Anomoeodus* species. The image on the left is the specimen collected by Dennis Magar from the Ozan Formation (Mid Campanian) along the NSR. The center image is *Anomoeodus (Pycnodus) phaseolus* from the Navesink Formation (Mid Maastrichtian) from New Jersey. The image on the right is the paratype of *Anomoeodus barberi* from the Marlbrook Marl (Upper Campanian) from Arkansas. In all images the anterior of the mouth is towards the top of the page. Notice the overall similarity of the three jaw sections. The differences are in the size and shape of the individual teeth. Does this constitute "normal" morphological variation within a single species? The specimen from the NSR is referred to as *Anomoeodus barberi* based on its closer proximity in space (Arkansas) and time (Mid to upper Campanian) compared to the New Jersey species.

Fig. 8 Three specimens of *Anomoeodus* from the Upper Cretaceous of North America.

Another interesting group of ray-finned bony fishes is the triggerfishes. So far, they are known only from microteeth recovered from the basal Ozan Formation (Plate 10, Figs.1 and 2) in the NSR study area. However, their teeth have also been discovered from the Maastrichtian Stage Kemp Clay in the South Sulphur River and from older Turonian Stage Kamp Ranch deposits in Collin County, Texas. All these are undescribed species. Bigger specimens collected in the field are commonly called "cat's claws" as they resemble the flattened claws of housecats.

Fossil triggerfishes belong to the family Eotrigonodontidae. This includes the genus *Eotrigonodon* from the Paleocene/Eocene of Italy, Angola and Egypt. The family also has numerous representatives from the late Cretaceous worldwide. These include *Pisduron* from Africa, *Indotrigonodon* from India, and *Stephanodus minimus* from the Maastrichtian Stage El Molino Formation in the Andes Mountains of Bolivia, South America. The latter form is described from isolated teeth that are indistinguishable from the microteeth from the basal Ozan Formation. *Stephanodus sp.* has also been recovered from the Maastrichtian age Navesink Formation in Monmouth County, New Jersey. The specimens from the Kamp Ranch Formation are possibly the oldest representatives known to date and await formal scientific description.

Modern day triggerfish have a worldwide distribution in tropical reef habitats. They are very beautiful with bright multi colored bodies. Adults range in size from 11 to 22 inches. They are very aggressive fish with incredibly powerful jaws that can chew through anything. Triggerfish can digest a variety of invertebrates including coral, sponges, molluscs and echinoderms. No doubt their ancestors from the NSR had a similar lifestyle and were a terror to the benthic community.

Table 6 is a listing of the marine sharks, rays and skates. Notice the decrease in diversity with increasing latitude. The Ozan and Mooreville faunas have far and away the most varied elasmobranch assemblages. Sharks and their kin are chondrichthyan fishes. Instead of a bony skeleton they have one composed of cartilage. This type of organic substance does not fossilize as well as bone. Usually teeth are the only common fossils in sharks and rays. Sharks shed their teeth through out their lives so one shark can leave behind hundreds of teeth. Occasionally shark vertebrae are found on the NSR (see Plate 17, Fig.15). Both micro and macro shark's teeth can be collected (Plate 17). It is not uncommon to find vertebrate material with the telltale scratch marks caused by the crow shark *Squalicorax* (see Chpt. 5).

Nicholls and Russell's 1990 paper was a summary of the paleobiogeography of the late Cretaceous Western Interior Sea (WIS) based on the vertebrate evidence. Previous workers had looked at various invertebrate groups including the ammonites, pelecypods, forams and others. All the research indicates that the Western Interior Sea can be subdivided into northern and southern subprovinces based on the distributions of the total marine biota. This is in turn a reflection of the paleoclimate during the early Campanian. Keep in mind that the Western Interior Sea stretched all the way from the present day Gulf of Mexico up to the Arctic Circle. The waterway itself was much shallower and narrower than either the Atlantic or Pacific Oceans. At its southern terminus it received subtropical waters from the Tethys Sea, and at its northern end there was an influx of colder, temperate waters from the Arctic Ocean. Somewhere in between these two temperature different water masses mixed. Just as warm air masses ride on top of colder ones, so too does warm water rise up and over colder waters. This leads to a temperature stratification of the water column in the Western Interior Sea.

The northern subprovince of the WIS is characterized by a low faunal diversity in all vertebrate groups. It is dominated by plesiosaurs, the shorebird *Hesperornis*, and the mosasaur *Platecarpus*. Note the lack of marine sharks and rays in Table 6. This is represented by the faunas at 1) Anderson River, Northwest Territories; 2) Pembina, Manitoba; 3) Sharon Springs, South Dakota; and 4) the Niobrara fauna of western Kansas. The southern subprovince of the WIS is dominated by sharks, turtles, and the mosasaur *Clidastes*. There is a higher diversity in all the vertebrate groups. Note the overall increase in fish diversity from north to south. The southern subprovince is represented by the localities 5) North Sulphur River, Texas; and 6) central Alabama.

Nicholls and Russell felt the boundary between the two faunal provinces was south of the Niobrara fauna and north of the Ozan fauna. Fig. 9 illustrates the approximate locations of the 6 localities listed above and in Tables 4-6.

So what does this mean to the average fossil collector who hunts the NSR? Use the vertebrate occurrence tables as a guide to potential fossil identification. There is a very good chance that any of the reported marine vertebrate fossils from the Niobrara of Kansas or the Mooreville of Alabama will also be discovered in the Ozan Formation someday. It is just a matter of time and diligent collecting. This also means it is very important to report any significant or unusual vertebrate finds to the proper authorities. Feel free to contact the FBI (Fossil Bureau of Investigation) branch of the Dallas Paleontological Society through the Dallas Museum of Natural History. We can put you in contact with a professional paleontologist or academic institution that might be interested in your discovery.

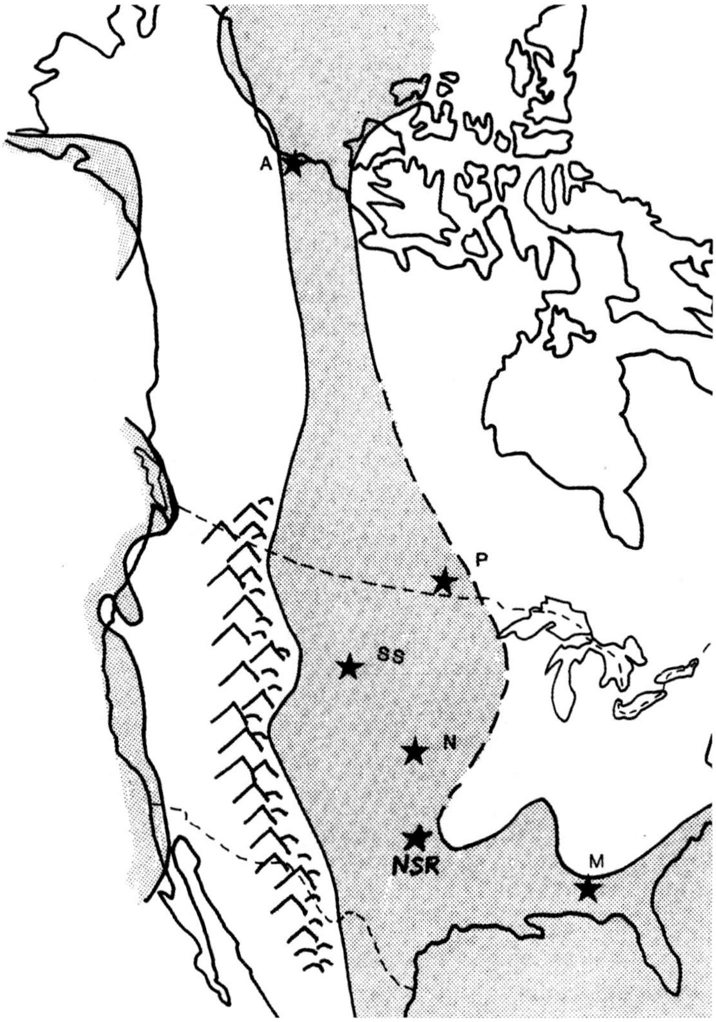

Fig. 9 Map of the 6 late Santonian through mid Campanian vertebrate localities listed in Tables 4-6. A=Anderson River, NWT, Canada; P=Pembina Member of the Pierre Shale, southern Manitoba; SS=Sharon Springs Member of the Pierre Shale, Wyoming; N=Smokey Hill Chalk of the Niobrara Formation, western Kansas; NSR=Ozan Formation, notheast Texas; M=Mooreville Chalk, central Alabama. WIS is stippled. Modified after Nicholls and Russell (1990).

Upper Cretaceous Vertebrate Occurrences – N. America

Table 4. – Marine Tetrapods

NW — SE

Genera	Anderson River – Pierre Sh.	Pembina Pierre Sh.	Sharon Springs – Pierre Sh.	W. Kansas - Niobrara Chalk	N. Sulphur R., N.E. Texas – Ozan Form.	Alabama – Mooreville Chalk
Mosasaurs						
Clidastes		X	X	X	X	X
Ectenosaurus				X		X
Globidens			X		X	X
Halisaurus				X	X	X
Hainosaurus		X				
Mosasaurus			X	X		X
Platecarpus		X	X	X	X	X
Plioplatecarpus	X					X
Polycotylus			X	X		
Prognathodon						X
Trinacromerum		X	X	X		X
Tylosaurus	X	X	X	X	X	X
Elasmosaurs						
Alzadasaurus		X	X	X		
Elasmosaurus		X		X		
Hydralmosaurus				X		
Styxosaurus			X	X		
Unidentified elasmosaurid	X					X
Turtles						
Archelon (Protostegid)				X		
Bothremys				X		X
Calcarichelys (Protostegid)				X		X
Chelosphargis (Protostegid)				X		X
Corsochelys (Dermochelid)						X
Ctenochelys (Chelonid)				X		X
Lophochelys (=Ctenochelys)				X		X
Porthochelys (Chelonid)				X		
Prionochelys (Chelonid)				X		X
Protostega (Protostegid)		X		X		X
Thinochelys (Chelonid)						X
Toxochelys (Chelonid)			X	X		X
Unidentified Chelonid turtle					X	X
Birds						
Baptornis				X		
Hesperornis	X	X	X	X		
Ichthyornis		X		X		X
Parahesperornis				X		
Pterosaurs						
Nyctosaurus				X		
Occidentalia				X		
Pteranodon			X	X		
Sternbergia				X		
Unidentified pterosaur						X

Upper Cretaceous Vertebrate Occurrences – N. America

Table 5. – Marine Fishes

Genera	NW Anderson River – Pierre Sh.	Pembina Pierre Sh.	Sharon Springs – Pierre Sh.	W. Kansas - Niobrara Chalk	N. Sulphur R., N.E. Texas – Ozan Form.	SE Alabama – Mooreville Chalk
Osteichthyans						
Albula						X
Anomoeodus (pycnodont)					X	X
Anquillavus				X		
Apsopelix			X	X		
Asarotus				X		
Bananogmius		X	X	X		X
Cimolichthys	X	X	X	X		X
Enchodus	X	X	X	X	X	X
Enchorhinus			X			
Ferrifrons				X		
Gillicus (ichthyodectid)			X	X	X	
Hadrodus (pycnodont)				X		X
Hoplopteryx					X	X
Ichthyodectes (ichthyodectid)		X	X	X		X
Kansius				X		
Laminospondylus				X		
Ledpisosteus				X		
Leptecodon				X		
Luxilites				X		
Martinichthys				X		
Micropycnodon (pycnodont)				X		
Moorevilla				X		X
Niobrara				X		
Omosoma				X		
Pachyrhizodous	X	X	X	X	X	X
Palelops				X		X
Paraliodesmus (amiid)				X		
Protosphyraena		X	X	X	X	X
Saurocephalus			X	X		X
Saurodon			X	X	X	X
Stephanodus		X	X	X	X	X
Stratodus		X	X	X		
Syntegmodus				X		
Urenchelys				X		
Xiphactinus (ichthyodectid)	X	X	X	X	X	X
Zanclites				X		

Upper Cretaceous Vertebrate Occurrences – N. America

Table 6. – Marine Sharks & Rays

NW → SE

Genera	Anderson River – Pierre Sh.	Pembina Pierre Sh.	Sharon Springs – Pierre Sh.	W. Kansas - Niobrara Chalk	N. Sulphur R., N.E. Texas – Ozan Form.	Alabama – Mooreville Chalk
Chondrichthyans – Sharks						
Anomotodon					X	X
Archeolamna					X	X
Cantioscyllium					X	X
Carcharias					X	
Chiloscyllium						X
Cretolamna		X	X	X	X	X
Cretoxyrhina				X		X
Edaphodon						X
Galeorhinus					X	
Ginglymostoma					X	
Heterodontus					X	
Hexanchus					X	
Lissodus					X	
Mustelus						X
Odontaspis			X	X	X	X
Paleogaleus					X	
Paranomotodon					X	
Pararhincodon					X	
Protolamna				X	X	
Psuedocorax				X	X	X
Ptychodus			X	X	X	X
Ptychotrygon					X	X
Rhincodon						X
Scapanorhynchus				X	X	X
Sclerorhynchus					X	X
Serratolamna					X	
Squalicorax		X	X	X	X	X
Squatina					X	
Chondrichthyans – Rays /Skates						
Brachyrhizodus					X	
Dasyatis					X	X
Hypolophus						
Ischyrhiza					X	
Myliobatis					X	
Onchosaurus					X	X
Psuedohypolophus						
Ptychtrygon					X	X
Rhinobatos					X	
Sclerorhynchus					X	

References

Bell, G.L., Jr. 1998. Middle Turonian (Cretaceous) Mosasauroids from Big Bend National Park, Texas. National Park Service Paleontological Research, 2:55-65.

Brasier, M.D. 1980. Microfossils.George Allen & Unwin, LTD. 194 p.

Case, G.R. 1982. A Pictorial Guide to Fossils. Van Nostrand Reinhold Company. 516 p.

Callaway, J.M. and Nicholls, E.L. 1997. Ancient Marine Reptiles. Academic Press. 502 p.

Dane, C.H., 1929, Upper Cretaceous formations of southwestern Arkansas. Arkansas Geological Survey, 1:215 p.

Finsley, C. 1989. A field guide to Fossils of Texas. Texas Monthly Press. 190 p.

Finsley, C. 1999. Discover Texas Dinosaurs. Gulf Publishing Company, 150 p.

Gayet, M., Sempere, T., Cappetta, H., Jaillard, E., and Levy, A. 1993. The presence of marine fossils from the late Cretaceous of the central Andes and it's paleogeographic consequences. Palaeogeography, Palaeoclimatology, Palaeoecology, 102:283-319.

Hill, R.T. 1894. Geology of parts of Texas, Indian Territory, and Arkansas adjacent to the Red River. Geologic Society of America Bulletin, 5:297-338.

Hirayama, R. 1997. "Distribution and diversity of Cretaceous chelonioids" in Ancient Marine Reptiles. Academic Press, 225-239.

Hussakof, Louis. 1947. A New Pycnodont Fish from the Cretaceous of Arkansas. Fieldiana. Chicago Natural History Mueum Press, vol. 10, no. 4, pp. 23-27.

Larson, N.L., Jorgensen, S.D., Farrar, R.A. and Larson, P.L. 1997. Ammonites and the other cephalopods of the Pierre Seaway. Geoscience Press,Inc., 148 p.

Maisey, J.G. 1996. Discovering Fossil Fishes. Henry Holt & Company, Inc., 224 p.

Nicholls, E.L. and Russell, A.P. 1990. Paleobiogeography of the Cretaceous Western Interior Seaway of North America: the vertebrate evidence. Paleogeography, Paleoclimatology, PaleoEcology, 79:149-179.

Romer, A.S. 1945. Vertebrate Paleontology. The University of Chicago Press (2nd ed.). 688 p.

Russell, D.A. 1993. Evolution of the Western Interior Basin - Vertebrates in the Cretaceous Western Interior Sea. Geological Association of Canada Special Paper, 39:665-680.

Schafer,W. 1972. Ecology and Paleoecology of Marine Environments. University of Chicago Press. 568 p.

Shiner, H.W. and Shrock, R.R. 1944. Index Fossils of North America. John Wiley & Sons, Inc. 838 p.

Stephenson,L.W. 1918. A contribution to the geology of northeastern Texas and southeastern Oklahoma. United States Geological Survey Professional Paper, 120-H:129-163.

Welton, B.J. and Farish, R.F. 1993. The collector's guide to fossil sharks and rays from the Cretaceous of Texas. BeforeTIME Publishing. 204 p.

As is the case in the oceans of today, the waters of the Campanian Stage of the Upper Cretaceous were patrolled by members of the class ***Chondrichthyes*** (translation = cartilage + fishes). Chondrichthyes is divided into two subclasses, ***Holocephali*** (Chimeras) and ***Elasmobranchii*** (Sharks and Rays). This chapter deals only with the sharks and rays since Chimera fossils are quite rare and have not, to the knowledge of the writer, been found in the strata exposed in the North Sulphur River. A glossary specific to shark and ray tooth terminology is included at the end of this chapter.

Aside from their teeth, which are fairly common in many areas, shark fossils are not easy to find or identify. The skeletal structure of sharks and rays is composed of cartilage rather than bone and is seldom completely preserved as fossils. Some areas of the skeletal structure, however, are partially calcified and are prone to being preserved in an identifiable state. The vertebral centrum is the segment most commonly calcified and preserved in a fossil state. The neural and hemal arches are somewhat less prone to calcification and therefore seldom preserved. See Fig. 1.

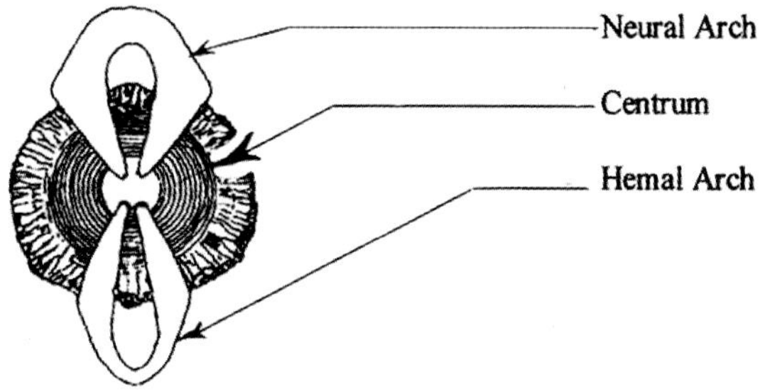

Fig. 1 Shark vertebra showing the centrum and the neural and hemal arches.

The vertebral centra of sharks can be distinguished from the vertebrae of bony fish by a number of characteristics. Shark vertebral centra are generally disk-shaped with both faces concave. The surfaces are fine-grained and similar to porcelain in appearance. The circumferencial surfaces usually exhibit two pairs of holes where the neural and hemal arches are anchored. Fish verterba, on the other hand, will show spiny processes where the neural and hemal arches and other bony projections were attached. The circumference of the vertebral centrum of a shark may be striated as in the Cretoxyrhind shark illustrated in Fig. 2, or smooth as in the Scyliorhinoid shark illustrated in Fig. 3.

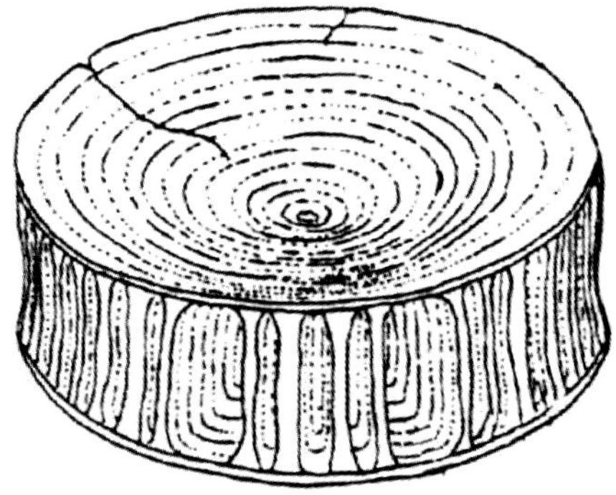

Fig. 2 Cretoxyrhinid-type Vertebral Centrum

Fig. 3 Scyliorhinoid-type Vertebral Centrum

The number of striations and the spaces between them is used to identify species with vertebral centra similar to Fig. 2. The number, size and spacing of the pores shown in Fig. 3 are the key to identifying sharks having the Scyliorhinoid-type of vertebral structure.

Fin spines and cephalic hooks may also be preserved and will be described later in the text where applicable to the genera with which they are associated.

Often overlooked, but still quite important in the fossil record of sharks and rays, are coprolites and feeding traces. Shark coprolites are characterized by their rolled, or spiral, appearance. This is caused by the spiral valve in the sphincter of sharks and rays. The unusually high incidence of well preserved shark coprolites, as opposed to those from bony fish, lies in the fact that shark dung is encased in a thin membrane when ejected by the shark. This membrane reduces the tendency of the fecal matter to be softened or dissolved by the seawater. The result is a neatly packaged candidate for fossilization. Fig. 4 shows a typical shark coprolite.

Fig. 4 Shark coprolite

Feeding traces consist mainly of scratches and gouges by the teeth of sharks in the bones of their prey. Less common are teeth still embedded in the fossilized bone. Both give valuable insight into the feeding preferences of various shark and ray genera. Figs. 5 and 6 show examples of damage to fossil bone inflicted by sharks in either killing or scavenging activities.

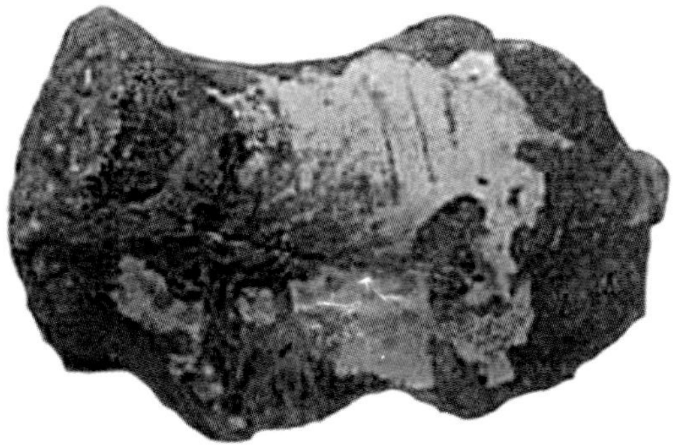

Fig. 5 Plesiosaur flipper bone showing teeth marks

Fig. 6 Mosasaur tail bone showing bite mark

The sharks and rays filled two ecological niches by acting in the roles of both predator and scavenger. Some preyed upon young or injured mosasaurs, plesiosaurs and ichthyosaurs and, in turn, were probably the prey of adult members of those families. Others acted as the sanitation engineers in their role as scavengers of carcasses of animals that died and settled to the bottom of the sea. Still others were eminently qualified to function in both roles as does the modern Tiger Shark.

The abundance of sharks' teeth in some strata is due, in part, to the fact that sharks lose and replace teeth throughout their lifetime. A modern Sand Tiger Shark, for example, will lose and replace a complete set of upper teeth every 8 days and a complete set of lower teeth once every 7 days, equating to the loss of many thousands of teeth over the duration of its lifetime. Many are lost in feeding which results in a fairly wide distribution across the floor of the shark's territory. Other teeth are lost during the mating ritual and will tend to represent fairly large accumulations of teeth from males of the same species. The reason that these teeth are usually from male sharks is that the males of most species grasp the fins of the females with their teeth during the mating act. Since this is probably a rather unpleasant sensation for the female, her struggles tend to dislodge quite a few teeth from the jaws of her would-be suitor. Since extant sharks tend to have specific breeding territories, it is reasonable to assume that their Cretaceous counterparts also frequented the same breeding areas year after year. The result is a tendency for rather large groupings of teeth to accumulate in relatively confined areas.

The third mechanism for the deposition of sharks' teeth is, as one would expect, the death of the animal. In the best case scenario, the teeth remain intact and maintain their relative jaw positions until covered with sediment and become fossilized. The more likely fate, however, is that the dead shark is scavenged by its own cousins and the teeth are swallowed or scattered.

The shark's mechanism of tooth replacement is quite interesting. Since the jaws are cartilage, rather than bone, the teeth are not tightly attached. As teeth are broken or lost in feeding or mating activity, the enamel and roots of new teeth are continuously forming on the innermost section of the jaw. As the enamel for a new tooth forms it displaces the previously formed enamel segment and forces it forward in the jaw. That segment, in turn, develops a root and pushes its predecessor forward to accommodate its additional space requirement. The process continues towards the outside of the jaw where the fully-formed teeth assume the position required to make them operational and, subsequently are lost due to crowding or one of the afore-mentioned activities. The "conveyor belt" tooth replacement mechanism is illustrated in Fig. 7.

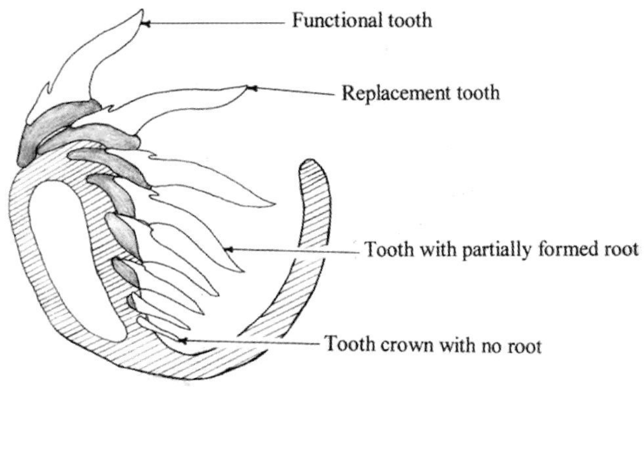

Fig. 7 Cross-section of jaw showing tooth growth and replacement mechanism.

A number of different orders represent sharks and rays in the Ozan formation. The most primitive of the sharks is the order **Hybodoniformes**. Included in this order are the genus *Ptychodus* and the genus *Lissodus.*

Examples of *Ptychodus latissimus* and *P. connleyi* have been reported from gravel bars in the North Sulphur River. To date, there has been only one report of the presence of *P. latissimus* in the NSR and its environs. It is generally accepted that *P. latissimus* represents one of the younger species of *Ptychodus*, with occurrences as late as the early Campanian Stage. This leads to the reasonable assumption that, if the examples found in the NSR were not inadvertantly carried into the river and lost by a fossil collector, they may be present in the basal Ozan Formation. Based on the size of *P. latissimus* teeth (which may easily exceed 30 mm), the fact that they have not been previously reported raises the probability they must be extremely rare in the Ozan Formation.

The teeth of *P. latissimus* are characterized by a moderately domed crown adorned with 5 or 6 massive transverse ridges exhibiting tapered ends and a slight curvature. The margins of the crown are quite wide and are adorned with a granular texture showing no distinct pattern. The root, which is overhung by the crown on all sides is rectangular in the basal view and lacks a nutrient groove.

The genus *Lissodus* is represented in the Ozan and Pecan Gap formations by at least 3 species, *Lissodus selachos, L. babulskii* and *L. sp.* (undescribed). The teeth of *Lissodus* normally do not exceed 4.5 millimeters in the greatest dimension, thus necessitating bulk sampling and screening to be employed for collecting. They exhibit a low crown with a prominent lingual protuberance as shown in Fig. 9. The teeth are rarely found with any portion of the root intact. All scale bars are 1 mm unless otherwise stated.

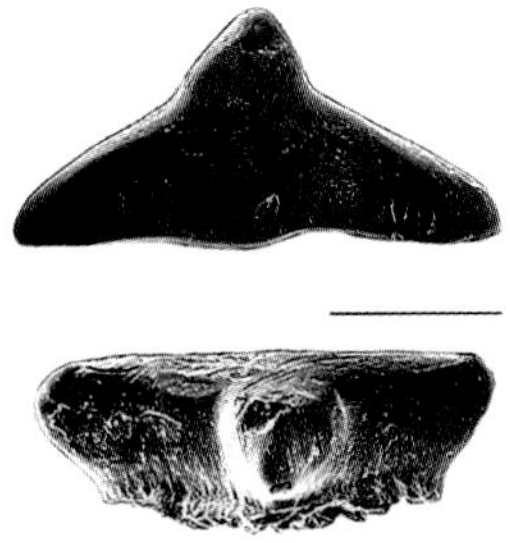

Fig. 8 Typical *Lissodus* tooth (3 views)

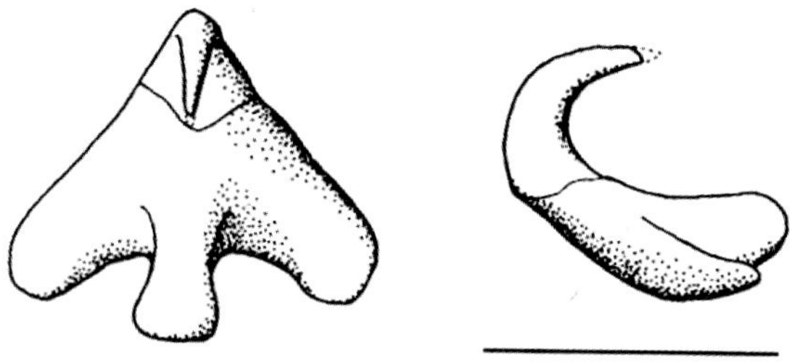

Fig. 9 Cephalic hook of *Lissodus sp.*

Lissodus is known to have cephalic hooks that appear to have been employed to clasp the female during the mating process. The hooks are characterized by their curved enameloid–covered cusp and a root–like basal section as illustrated in Fig. 9. The curved section often displays longitudinal striations. *Lissodus* also possessed pointed spines in front of the leading edge of the dorsal fin. These spines are usually covered with tubercles on the leading edge and exhibit at least one row of denticles along the trailing edge. See Fig. 10.

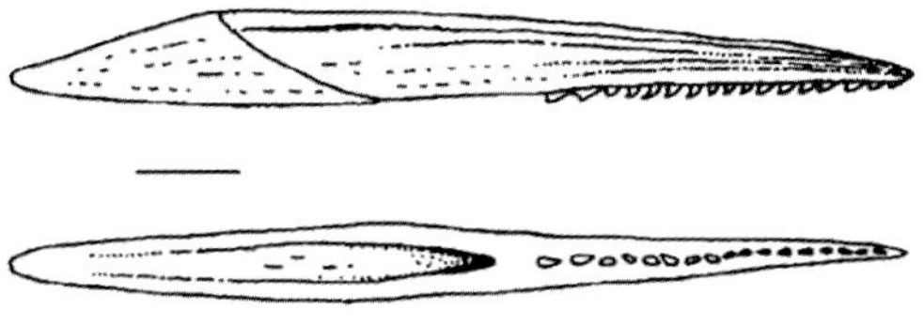

Fig. 10 Dorsal spine of *Lissodus sp.*

Lissodus is but one of a number of shark genera represented in the North Sulphur River by micro-sized teeth. Some are from the Cat Shark family (Scyliorhinidae), whose members were probably less than ½ meter in length. Other microteeth, surprisingly enough, can be from fairly large individuals. While the oral teeth of the sawfish *Ischyrhiza mira* are only about 5 to 6 millimeters in the greatest dimension, the shark itself may well have exceeded 5 meters in length.

One of the most interesting families of shark equipped with micro-sized teeth is that known as **Schlerorhynchidae**. Schlerorhynchidae or more commonly, Schlerorhynchids, are members of the order **Rajiformes** (rays). While the family contains a great number of species, the teeth of all Schlerorhynchids are markedly similar. Despite this similarity, some show a tendency to be utilized for crushing food while others exhibit the attributes of clutching dentitions.

A member of the Schlerorhynchid family equipped with a crushing dentition and commonly encountered in both the Ozan and the Wolf City Formations is *Ptychotrygon triangularis.* This genus is not known, with any degree of certainty, to have been endowed with an elongated snout containing rostral teeth as in the sawfishes, but is considered by some to have had a bill not unlike that of a marlin. Both viewpoints are arguable. Four views of a *P. triangularis* oral tooth are shown in Fig. 11.

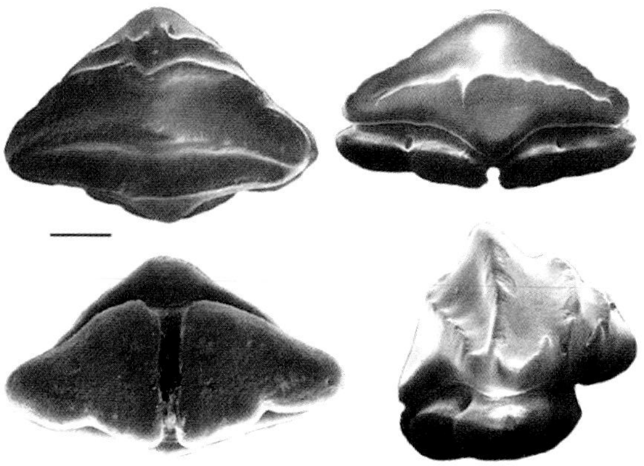

Fig. 11 *Ptychotrygon triangularis*

The sawfish, are represented in the Ozan Formation in the North Sulphur River by the genera *Ischyrhiza* and *Sclerorhynchus*, both of which are extinct. They do, however, bear strong resemblance to extant species and probably were similar in life style and feeding methods. As with sharks the most commonly encountered fossil remains are the teeth. In the case of the sawfish, this includes teeth (or spines) which were located along both sides of the long, flattened bill or rostrum, as well as oral dentition. Modern sawfish utilize these rostral teeth by swimming into a group of smaller fishes and swinging their head (and rostrum) to impale a victim on the rostral teeth. They then have been observed to drop to the sea floor and gently dislodge the prey fish from their rostral teeth. All the while using the rostrum to pin the prey down. They then simply swim forward, while still holding the prey down, and engulf the prey in their mouth. In other cases, they simply cripple members in a school of fish and leisurely feed on the wounded individuals.

Ischyrhiza avonicola oral teeth (Fig. 12) bear a very strong resemblance to those of *Ptychotrygon triangularis*, while those of *Ischyrhiza mira* show more of an affinity for clutching rather than crushing as shown in Fig. 13.

Sclerorhynchus sp. Oral teeth bear a resemblance to those of the genus Ischyrhiza , but are readily distinguishable by the radiating ridges on the top of the crown as evidenced in Fig. 14. The oral teeth of *Ischyrhiza* and *Sclerorhynchus* tend to be quite small, ranging from 1.5 millimeters for *I. avonicola* to 6 millimeters for *I. mira*. The oral teeth of *Sclerorhynchus sp.* are very slightly larger than those of *I. avonicola*.

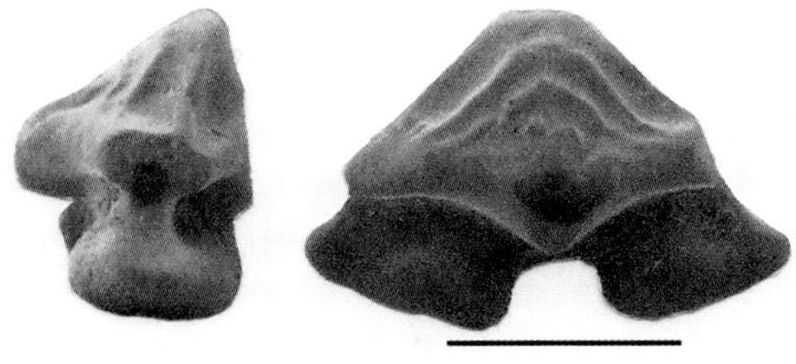

Fig. 12 *Ischyrhiza avonicola* oral tooth

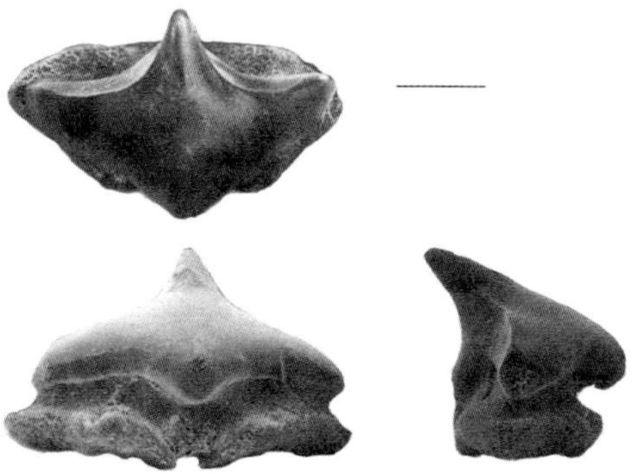

Fig. 13 *Ischyrhiza mira* oral tooth

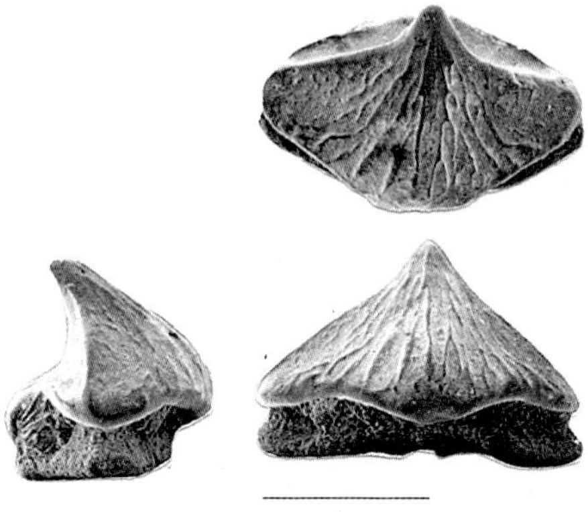

Fig. 14 *Sclerorhynchus sp.* Oral tooth

The difference between the sawfish genera and species is most apparent in the rostral teeth. The rostrals of *Ischyrhiza avonicola*, with a maximum height of 4 millimeters, have crowns that are posteriorly inclined and are less than half the total height of the tooth. The anterior surface is convex and exhibits a cutting ridge running from the crown base to the tip. Short longitudinal ridges, or folds surround the enameloid crown base. The peduncle (root) is widely splayed. See Fig. 15.

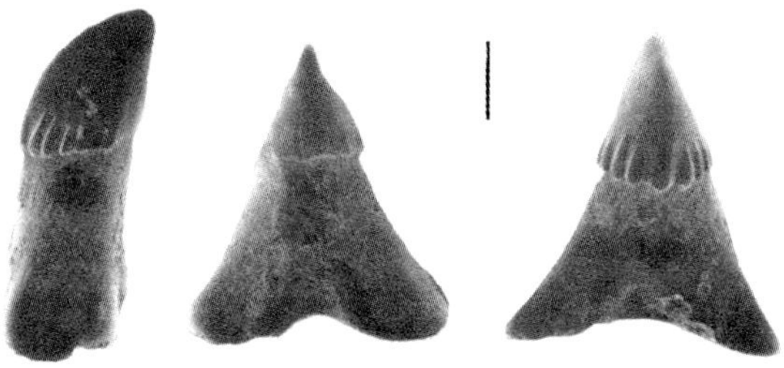

Fig. 15 *Ischyrhiza avonicola* Rostral tooth

The rostrals of *Ischyrhiza mira*, on the other hand, are large and blade-like with a tall, massive peduncle (root) containing a longitudinal basal furrow and scalloped root lobes. The crowns have a continuous cutting edge that encompasses both the anterior and posterior surfaces of the crown. The overall length of the rostrals may exceed 50 millimeters. See Fig. 16.

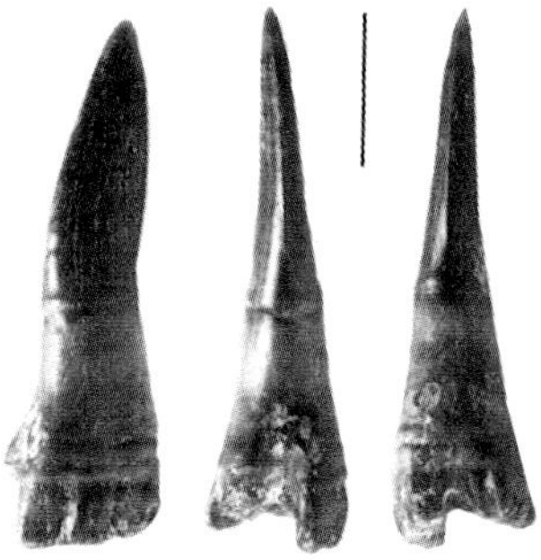

Fig. 16 *Ischyrhiza mira* Rostral teeth

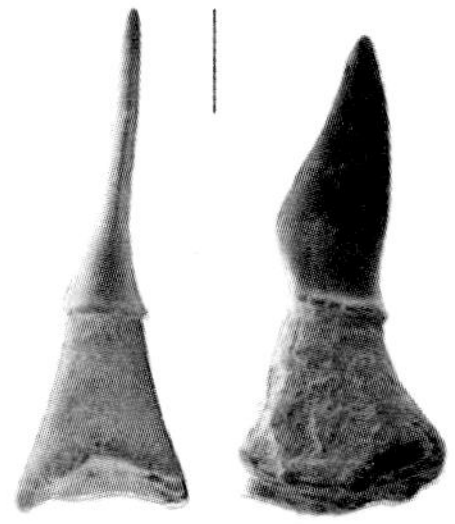

Fig. 17 *Sclerorhynchus sp.* Rostral tooth

The rostral teeth of *Sclerorhynchus* fall between the size of *I. mira* and *I. avonicola*. They are readily distinguishable from *I. mira* by the size difference and by the shape of the blade. The blade of Sclerorhynchus, when viewed in profile, has a distinct bulge on the anterior edge. This bulge in some cases may actually form a small hook. The peduncle is less than half the overall height of the tooth and has a concavity at the base running anteroposteriorly. See Fig. 17.

Other members of the order Rajiformes found in the Ozan are the guitarfishes or **Rhinobatidae**. Their teeth are quite small, measuring 1.5 to 2 millimeters, and are characterized by a long protuberance on the foot of the lingual crown. This protuberance is apparent on the image on the left side of Fig. 18. There is a strong likelihood that a number of species of *Rhinobatos* is represented in the NSR.

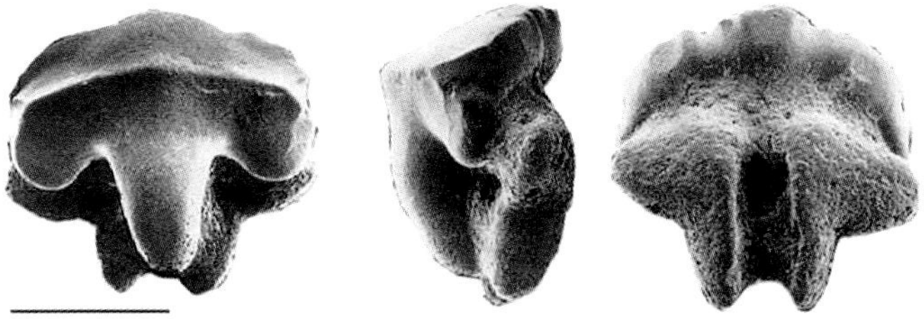

Fig. 18 *Rhinobatos sp.*

The order **Myliobatiformes** includes the eagle ray, the cownosed ray and the stingray. It is, in part, represented in the Ozan by a number of species included in the genus *Dasyatis*. The teeth of *Dasyatis* are quite small, seldom exceeding 3 millimeters in the greatest dimension.

The teeth of males and females differ greatly within a species, creating some confusion when attempts are made to describe and name individual species. Suffice to say, there are a number of species within the genus *Dasyatis* that will remain unnamed until study can be done. The tooth pictured in Fig. 19 represents a species as yet unnamed, similar to, but differing slightly from a later Texas species known as *Dasyatis commercensis* (Case and Cappetta, 1996).

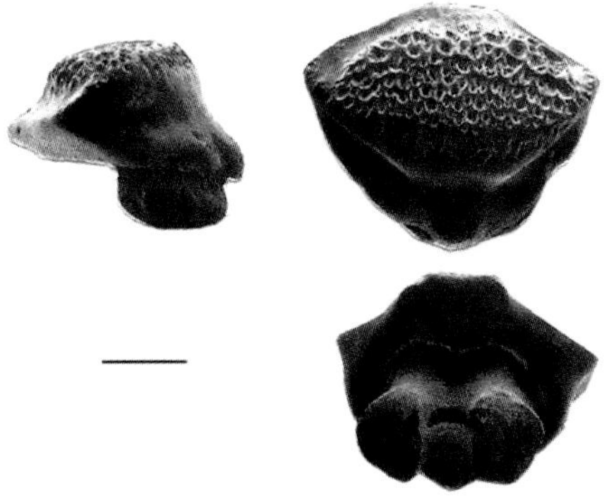

Fig. 19 *Dasyatis sp.* (Scale line =0.5 mm)

Fig. 20 *Myliobatis sp.*- Occlusal and basal views

The rarest of the Texas Cretaceous Myliobatiformes is the genus *Myliobatis*. The teeth of *Myliobatis* are easily recognized by their relatively flat, unornamented, occlusal crown surfaces as well as the numerous nutrient grooves on the root. The mesodistal dimension is much greater than the labiolingual dimension, with the mesial and distal ends tending to have a triangular appearance as shown in Fig. 20. The overall width of the tooth can be in excess of 11 millimeters. By the shape of their teeth, it is reasonable to assume the *Myliobatis* found in the Ozan Formation was a benthic ray with the same feeding habits as its modern descendants.

Another of the rare members of the shark family found in the Ozan is the **Heterodontiformes** which include the genus *Heterodontus. Heterodontus* exists today as a small, bottom-dwelling shark that rarely exceeds one meter in length. It inhabits shallow tropical and temperate waters around the world. It is characterized by its short, robust head with prominent ridges above the eyes and small mouth. The two dorsal fins are preceded by sharp spines. The body is covered with fairly large placoid scales bearing a distinctive ornamentation resembling a Maltese cross.

Heterodontus is represented in the Texas Cretaceous by at least one species (*Heterodontus* cf. *canaliculatus*). The anterior teeth, as illustrated in Fig. 21, are seldom found intact in Texas. As a result, it is difficult to assign the teeth found in the Ozan to any specific species. Typically, the anterior teeth have a tall cusp that tends to be rounded at the apex. It is flanked by one pair of robust cusplets. The crown faces are usually smooth. There is a nearly horizontal lingual protuberance on the crown that overhangs the root. The root lobes form a U-or "V", with the opening on the labial side of the tooth. The root attachment surface is perforated by a number of foramen.

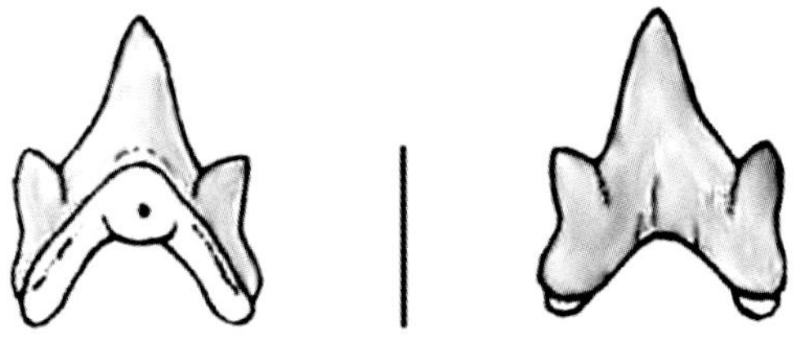

Fig. 21 *Heterodontus sp.* Anterior tooth
Scale line = 1 mm

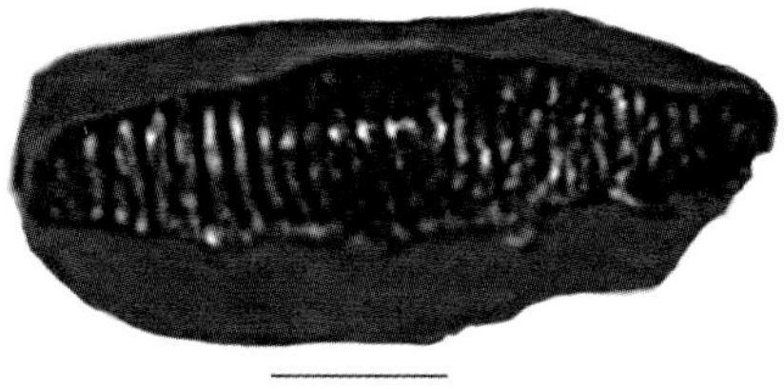

Fig. 22 *Heterodontus sp.* Posterior tooth

Heterodontus is unique in that the anterior teeth exhibit the characteristics of a clutching dentition, while the posterior files represent a crushing dentition (see Fig. 22). These teeth bear a slight resemblance to the teeth of *Ptychodus*, but are more symetrical. They also exhibit a greater mesodistal dimension relative to their labiolingual dimension. A median transverse ridge is present on all posterior teeth. Identification of both genus and species is complicated by the fact that the shape of *Heterodontus* teeth changes throughout the life of the individual.

Another interesting order of sharks to be found in the Ozan is the **Orectolobiformes**. They consist of a diverse group of predominantly small, shallow water, bottom-dwelling sharks. The exception to this rule is the Whale shark which is the world's largest living shark. Most Orectolobiforms are characterized by a blunt snout, a narrow mouth equipped with numerous teeth, two dorsal fins and a very long tail. The teeth tend to be quite small with few species possessing teeth in excess of 3 millimeters in the maximum dimension.

The Orectolobiform species most often encountered in the Ozan is *Cantioscyllium meyeri.* The teeth exhibit short, stout cusps flanked by from one to three pairs of cusplets. The lower surface of the labial face of the crown is covered with prominent ridges, or folds, that do not reach the apex of the cusp as shown in Fig. 23. The root is triangular in outline with well-defined lobes separated by a U-shaped groove as evident in Fig. 24.

Fig. 23 *Cantioscyllium meyeri*
Labial view

Fig. 24 *Cantioscyllium meyeri*
Basal view

An extremely rare Orectolobiform, *Pararhincodon groessenssi*, occurs in the Ozan. It is characterized by extremely small teeth (typically 0.5 to 1.2 millimeters in the maximum dimension). The cusp is needle-like and is usually flanked by a low blade-like cusplet on the mesial side. A distal cusplet is usually well developed. A continuous cutting edge runs across the cusplets and cusp. The root is bilobate, with a large lobe on the mesial side and a small lobe on the distal side. The lobes are separated by a nutrient groove. See Fig. 25.

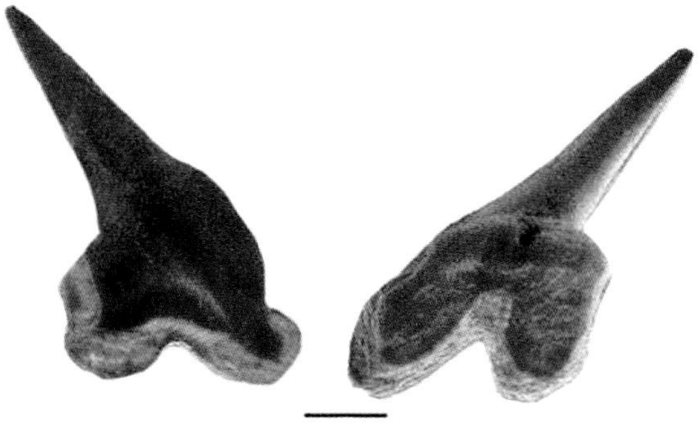

Fig. 25 *Pararhincodon groessenssi*
Scale line = 0.2 mm

The family **Squatinidae** (angel sharks) includes the genus *Squatina.* This genus has an extremely long fossil record, extending from the Jurassic to the present. Although complete skeletons from the Jurassic have been found, the fossil record of *Squatina* consists of isolated teeth and scales throughout the Cretaceous and Tertiary. Due to the similarity of teeth throughout its long history, it is extremely difficult to identify any given species.

Superficially, *Squatina* looks more like a ray than what we normally consider the "classic" shark body plan. The body is dorsoventrally flattened with broad pectoral fins. The dorsal fins are tiny and located well back on the body.

Teeth of *Squatina* are quite rare in Texas, but can be found in the basal Ozan in the Ladonia area. They are small (typically 3 to 5 mm) and easily confused with the oral teeth of *Ischyrhiza mira* when viewed from above. The difference between the two is readily apparent in the basal view, with Squatina possessing a broad, weakly concave attachment surface. *I. mira* exhibits a diamond-shaped root, bissected by a deep nutrient groove when viewed in the basal aspect.

The species most likely to be encountered in strata bordering the NSR is *Squatina hassei,* illustrated in Fig. 26.

Fig. 26 *Squatina hassei*

An interesting family of sharks known as **Triakidae** first appear in the early Campanian and are represented in the Ozan in numerous locations along the NSR by the species *Palaeogaleous navarroensis*. This species was named by Case and Cappetta based on teeth found in Navarro Group sediments in Hunt County, Texas. Only two other Campanian species are known, both being from Europe.

The teeth of *Palaoegaleous* are small (3 – 3.5 mm) and are characterised by a very robust crown and root. The cusp is thick and is flanked by cusplets, the number and position of which varies with location in the dental series. The basal ledge of the crown overhangs the root on the labial face and is ornamented with short ridges that extend about halfway up the crown. The root is massive and consists of two triangular lobes, separated by a wide and deep nutrient groove. A continuous cutting edge runs along the cusplets (when present) and primary cusp. *P. navarroensis* is pictured in Fig. 27.

Fig. 27 *Palaeogaleous navarroensis*
Lingual and labial views

A fascinating and diverse order of sharks represented in the stata bordering the North Sulphur river is that of **Lamniformes.** The order includes the family **Anacoracidae**, of which at least three species are represented in the Ozan strata are generally believed to belong. There has been considerable debate over the years questioning the relationships of the Anacoracidae with respect to Order and Family.One of the more controversial members of the family is *Pseudocorax granti.* It is particularly abundant in strata from the lower Ozan.

P. granti appears to have been a inhabitant of warm, shallow seas and probably acted as both predator and scavenger. Its teeth exhibit a smooth cutting edge that would appear to be fairly effective in dispatching soft-bodied prey. The cusps are quite thin and have a fragile appearance. The anterior teeth are short and narrow with erect crowns and are quite easily mistaken for teeth of *Paranamotodon* (mentioned later in this text). The lateral teeth have a well defined distal blade as opposed to the mesial blade which almost appears to be an extension of the cusp. See Fig. 28.

The root is strongly bilobate and differs from that of other Anacoracidae in that it has a pronounced nutrient groove. It is also fairly small, with a maximum dimension of about 9 mm.

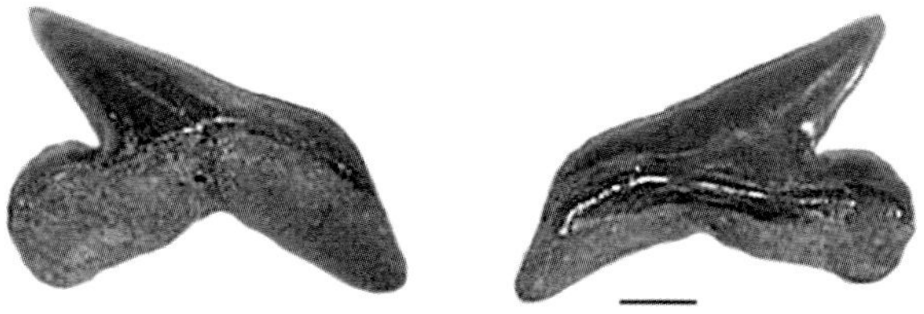

Fig. 28 *Pseudocorax granti*
Lingual and labial views

Another genus belonging to the family Anacoracidae and represented in the Ozan strata exposed by the North Sulphur River is *Squalicorax*. The teeth of this genus resemble those of the tiger shark that roams the modern seas. The tiger shark is both a predator and scavenger, a trait probably shared by *Squalicorax.* There is also a strong possibility that *Squalicorax* prowled deep waters during the day and moved into shallow offshore and estuarine waters after dusk, as does the tiger shark.

It is interesting to note that teeth from *Squalicorax* have been found in association with the skeletal remains of large sharks, such as *Cretoxyrhina mantelli.* In at least one instance, the difference in the size of the teeth of the *Cretoxyrhina* versus those of the *Squalicorax* indicate that *Squalicorax* would have constituted little more than a snack for the *Cretoxyrhina*. The conclusion to be drawn from this association seems to be evidence of scavenging of a *Cretoxyrhina* carcass by *Squalicorax.*

Two species of *Squalicorax* can be found in the environs of the river. The older of the two is *Squalicorax kaupi* and can be found in all the Cretaceous strata along the NSR. *S. kaupi* is easily distinguished from *Pseudocorax granti* by its much more robust appearance and the absence of a nutrient groove on the root. It is considerably larger than *P. granti*, with a maximum dimension of 15 millimeters.

S. kaupi is characterized by its high crown bearing a convex serrated cutting edge on the mesial side. The distal side of the cusp has a nearly straight serrated cutting edge that is nearly vertical. A low distal blade is present. The juncture of the cusp and blade forms an obtuse, but discernable angle. As is the case with the cusp, the blade is serrated.

The root is almost as high as the crown on the lingual face, with a nearly non-existent lingual protuberance. The labial side of the root is considerably lower than the lingual side. The basal surface of the root has a profile ranging from moderately to extremely concave, with the root lobes rounded as shown in Fig. 29.

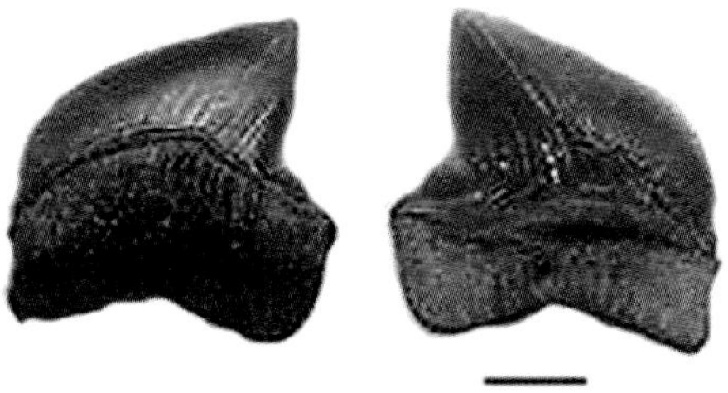

Fig. 29 *Squalicorax kaupi*
Labial and lingual views

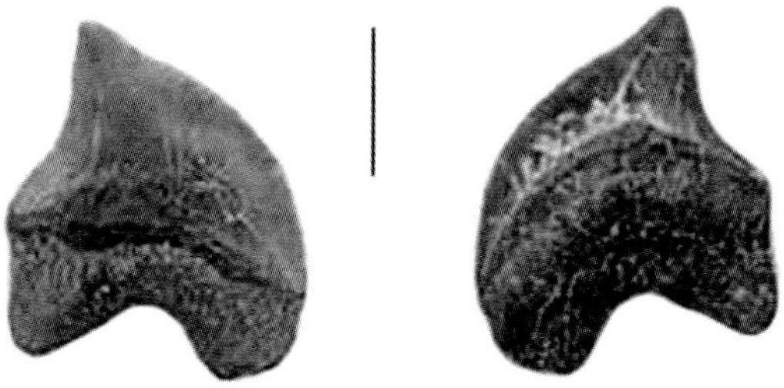

Fig. 30 *Squalicorax pristodontus*
Labial and lingual views

The other *Squalicorax* species to be found in the NSR is *Squalicorax pristodontis*. This is the largest member of the genus *Squalicorax*, with teeth measuring in excess of 29 millimeters.
The cusp and root of *S. pristodontis* are relatively thin for a tooth of its size, and both exhibit a strongly concave profile when viewed mesodistally. The profile of the basal surface of the root is quite similar to that of *S. kaupi*. As in the case of *S. kaupi*, the lingual face of the root is very high, sometimes exceeding the height of the crown. The labial face of the root is much lower than the lingual face.

The crown is fairly erect and has a broadly convex mesial cutting edge which sometimes recurves slightly near the tip of the crown. The distal cutting edge is usually straight and nearly vertical. It intersects with a distal blade at a very obtuse angle as shown in Fig. 30. In some instances, the juncture of the crown and the blade is not discernible. The crown and blade are always coarsely serrated.

Squalicorax pristodontis can be easily confused with S. *kaupi*, especially in the case of larger teeth. While the generally accepted rule of thumb states that teeth with a distinct angular, or notched blade-crown junction are *S. kaupi* and those with an obtuse junction are *S. pristodontis*, the differences are sometimes to small to provide positive identification.

Another interesting branch of the order Lamniformes is the family **Cretoxyrhinidae**. This family is represented in the NSR by a number by a number of genera and species.

The first belongs to the genus *Cretolamna* and is represented by the species *Cretolamna appendiculata*. *C. appendiculata* is an especially tenacious species, having survived from the Albian and into the Eocene Epoch.

The anterior teeth of *C. appendiculata* are moderately large, attaining a maximum dimension in excess of 26 millimeters. The tall, narrow cusps have completely smooth labial and lingual faces. The labial face is nearly flat as opposed to the lingual face which is strongly convex. The cusp is weakly sigmoid when viewed in profile, and is flanked by a pair of closely attached triangular cusplets. The roots are strongly bilobate with a flat attachment surface and a weak lingual protuberance. The lobes are separated by a deep U-shaped notch. There is no nutrient groove. See Fig. 31.

The lateral teeth have cusps that are more labiolingually compressed than the anteriors and are distally inclined as shown in Fig. 32. The labial surface is flat and the lingual is broadly convex. The triangular cusplets are closely attached. The root lobes are separated by a much shallower notch than is present in the anteriors. The basal surfaces of the root lobes may be broadly rounded or flattened depending upon the dental position. The lingual protuberance on the attachment surface is small to diminutive. No nutrient groove is present and both crown faces are smooth.

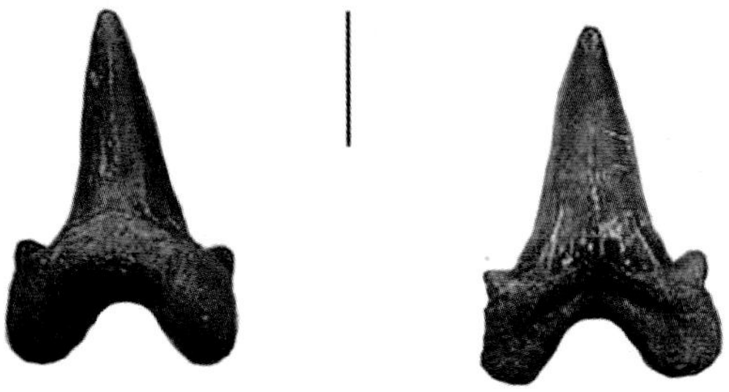

Fig. 31 *Cretolamna appendiculata*
Anterior tooth

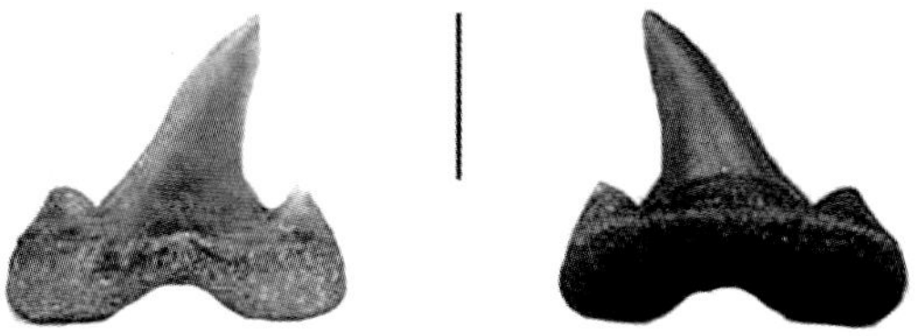

Fig. 32 *Cretolamna appendiculata*
Anterolateral tooth

The second Cretoxyrhinid found in the NSR is *Archeolamna koepingensis*. *Archeolamna* was, until recently, felt to belong to the genera *Cretodus*. The feature that separates it from *Cretodus* also makes it easy to confuse with *Cretolamna appendiculata*. *Cretodus* is characterized by a cusp that is heavily striated on the lingual face just above the root. The cusp of *Archeolamna* is completely smooth. This characteristic is shared by *Cretolamna*, as is a similarity in the shape of the cusp and cusplets. The notch between the root lobes, however is more pronounced in *Archeolamna* than in *Cretolamna*. In addition, the lingual protuberance on the root of *Archeolamna* is highly prominent and actually forms a shelf at the base of the crown. *Archeolamna koepingensis* reaches a maximum size of around 30 millimeters. An anterior tooth is shown in Fig. 33.

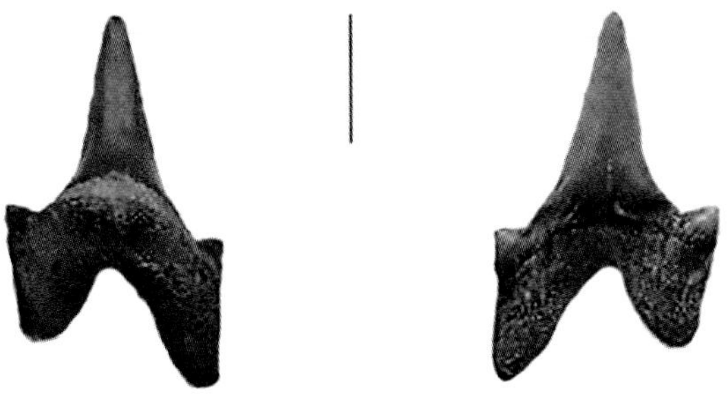

Fig. 33 *Archeolamna koepingensis*

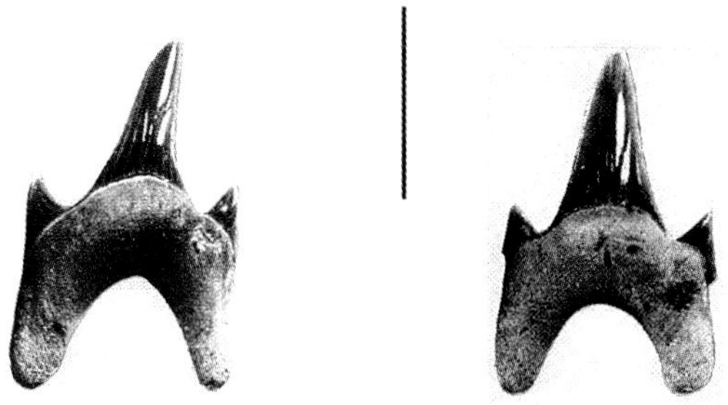

Fig. 34 *Protolamna sp.*

The third member of the Family Cretoxyrhinidae to be found in the NSR is the genus *Protolamana*. An undescribed species of *Protolamna* is quite commonly encountered in the basal Ozan. The teeth are quite small, averaging about 6 millimeters in height. They are characterized by a tall, narrow cusp flanked by a pair of large triangular cusplets. The cusp and cusplets are heavily striated at the juncture of crown and root. The striations extend about half way up the height of the cusp and cusplets. The lingual face of the cusp is also striated.

The root is strongly bilobate with the lobes separated by a deep V-shaped notch. As in the case of *Archeolamna*, the lingual face of the root exhibits a prominent centrally located protuberance. The protuberance forms a shelf at the base of the crown. See fig. 34.

A Lamniform shark that is infrequently encountered in the NSR is *Serratalamna serrata*. It is the sole representative of the family **Serratolamnidae** to be found in the Cretaceous of Texas. It is uncommon in the strata bordering the upper drainage of the NSR, but does make its appearance in strata from the late Campanian downstream.

Cusps of the teeth from all dental positions are labiolingually compressed, with a weakly convex labial face and a nearly flat lingual face. The cusp are broad-based and flanked by multiple pairs of asymmetrical diverging cusplets. There is often one more cusplet on the distal side of the cusp than on the mesial side.

The root lobes are separated by a shallow V-shaped notch and are weakly bilobate and horizontally expanded. A nutrient groove is present on the labial face. The feature that distinguishes *Serratalamna serrata* from all other lamniforms is the asymmetrical structure of both the crown and root in combination with the flattened crown and divergent cusplets. An anterolateral tooth is shown in Fig. 35.

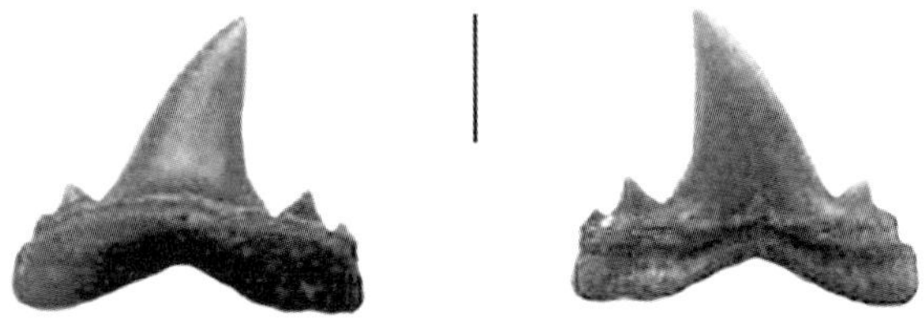

Fig. 35 *Serratalamna serrata*
Lingual and labial views

Fig. 36 *Paranamotodon sp.*
Lingual and labial views

A shark found in the Ozan, and generally considered to be a member of the family **Alopiidae** (thresher sharks) is *Paranamotodon sp.*. This is a somewhat problematical genus with inclusion in the family Alopiidae based on dental similarities to a previously identified extinct thresher shark.

Paranamotodon is characterized by a high cusp with smooth labial and lingual faces. The cusp has broad labial and lingual shoulders that are almost horizontal in the lateral teeth. The cutting edge is continuous across the shoulders and cusp. In some instances, an incipient cusplet is present at the outer extremities of the shoulders.

The root is bilobate with the lobes separated by a shallow U-shaped notch. A nutrient groove is always present and is located on the lingual protuberance as shown in Fig. 36.

The largest of the sharks to be found in the environs of the river are assigned to the family **Mitsukurinidae** which includes the extant goblin shark. The goblin shark is a bizarre creature with a thin, elongated snout, greatly protruding jaws, very long tail and eyes located over the corners of the mouth. The overall length of some individuals exceeds 11 feet. Goblin sharks live on or near the bottom at depths in excess of 1500 feet. The fossil evidence suggests that they have remained in this habit for at least 60 million years.

The family is represented in the NSR by *Scapanorhynchus texanus*. The teeth of *S. texanus* are very similar to those of the extant sand tiger shark and to those of Odontaspids dating from the Paleocene and throughout the Tertiary. Although it is quite easy to mistake the teeth of Upper Cretaceous Odontaspids for juvenile *Scapanorhynchus texanus*, the fully grown adult S. *texanus* stands out when compared to its peers.

The anterior teeth are quite large and may attain lengths in excess of 48 millimeters. The long, narrow, cusp has strong striations extending from the base of the crown almost to the tip. The cusp narrows at a point about one third of the way up from the base. The lingual crown face is strongly convex whereas the labial face is nearly flat. The cusp is strongly sigmoid in profile. Cusplets are normally absent, but one diminutive pair may occur on isolated individuals. A narrow dental band exists between the crown and the root. The dental band usually exhibits vertical striations in both the labial and lingual aspects. The striated dental band is often an important feature to be used in the identification of the teeth of this species.

The root of the anteriors is strongly bilobate, as shown in Fig. 37. A strong lingual protuberance, bisected by a deep nutrient groove, forms a ledge at the base of the crown.

The lateral teeth possess mesodistally compressed cusps, with the lingual face being broadly convex and the labial face nearly flat. The crown faces range from weakly striated to smooth. The crown bases are mesodistally expanded, as is evident in Fig. 38, and are flanked by one or two pairs of cusplets. The bilobate root is mesodistally expanded and exhibits a prominent nutrient groove located in the center of a weak lingual protuberance. The root lobes are normally broadly rounded.

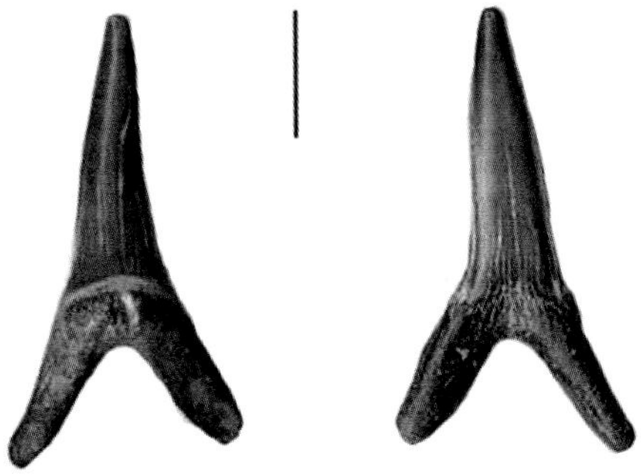

Fig. 37 *Scapanorhynchus texanus*
Anterior tooth

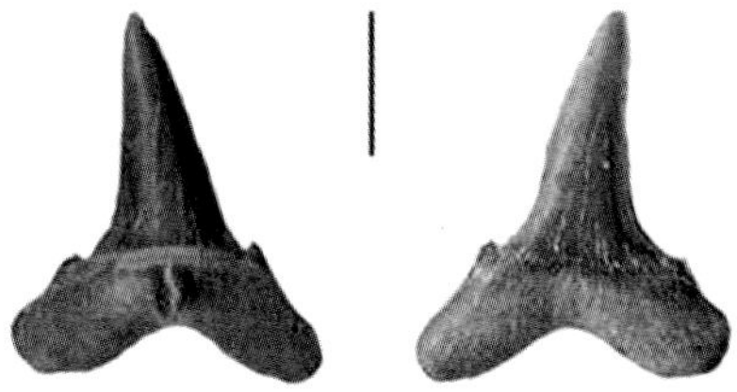

Fig. 38 *Scapanorhynchus texanus*
Lateral tooth

At least three small **Odontaspids**, notably the *Charcharius* genera, are present in the strata bordering the NSR. While they bear strong affinities to previously described species, much more study is required to positively place them as regards genus and species.

A very notable presence in the Pecan Gap formation exposed on the NSR is that of *Hexanchus microdon*, an example of which is shown in Fig. 39. The modern members of the Hexanchus family are commonly referred to as "six-gill sharks". They inhabit the cold water near the bottom at depths of as much as 650 feet.

Hexanchus microdon is characterized by lateral teeth possessing a saw-like crown with numerous cusplets, supported by a horizontally expanded tabular root. The primary cusp, which is only marginally larger than the cusplets is serrated on the mesial side. The cusplets are not serrated. The incidence of *Hexanchus microdon* teeth is extremely rare in Texas.

Fig. 39 *Hexanchus microdon*

Glossary of Terms

Aff. - Affinity. Used to indicate a species closely related to, but not identical to, a previously described species.

anterior- Front

anterolaterals- A rowgroup of teeth in which the similarities between anterior and lateral teeth make distinguishing between the two nearly impossible.

apical- The view seen when looking at the tip of the crown of the tooth from above.

Attachment surface- The portion of the root that seats in the dental membrane against the jaw surface.

basal- The view seen when looking at the root of the tooth from below.

basal ledge- The ledge formed by the expansion of the crown foot above the root.

bilobate- Root having 2 lobes, distal and mesial.

cf.- Comparable to a described species.

coprolite- Fossilized excrement. The non-digestible portion of food passed through the digestive tract of an animal.

crown- The enameloid-covered portion of a tooth.

cusp- The major crown projection of a tooth.

cusplet- A minor crown projection flanking the crown of a tooth.

distal- The opposite of mesial. Toward the hinge of the jaw.

dorsal- Toward the top.

dorsoventrally- From top to bottom.

foramen- Hole.

labial- The face of a tooth facing the outside of the mouth.

labiolingual- The orientation of a tooth relative to the outside and the inside of the mouth.

lingual- The face of the tooth facing the tongue.

mesial- Opposite of distal. Toward the front of the jaw.

mesodistal- Referencing front to back of the jaw.

NSR- North Sulphur River.

nutrient groove- A groove separating the root lobes on the basal, or lingual root face.

occlusal- Same as apical.

oral- Teeth found in the mouth.

rostral- Spine-like teeth found along the edges of the flattened rostrum of sawfishes.

rostrum- Elongated snout of sawfishes and other billfish.

sigmoid- Having a shape resembling the letter "S".

Chapter 5 References

Cappetta, H. and Case, G.R. ca. 1970. Les Selaciens de Groupe Monmouth (Campanien – Maastrichtien) de New Jersey. 105 p.

Case, G.R. 1973. Fossil Sharks: A pictorial Review. 64 p.

Case, G.R. and Cappetta, H. 1996. A New Selachian Fauna from the Late Maastrichtian of Texas. Muncher Geowiss. Abh.

Case, G.R. and Schwimmer, D.R. 1988. Late Cretaceous Fish from the Blufftown Formation (Campanian) In Western Georgia. Journal of Paleontology, vol. 62 (2). pp. 290 -301

Kent, B.W. 1994. Fossil Sharks of the Chesapeake Bay Region. 136 p.

Lauginiger, E.M. and Hartstein, E.F. 1983. A Guide to Fossil Sharks, Skates and Rays From The Chesapeake and Delaware Canal Area, Delaware. State of Delaware, University of Delaware, Delaware Geological Survey Open File Report No. 21. 65 p.

Meyer, R.L. 1970. Late Cretaceous Elasmobranchs from the Mississippi and East Texas Embayment of the Gulf Coastal Plain. Unpublished doctoral dissertation. 419 p.

Siverson, M. 1992. Biology, Dental Morphology and Taxonomy of Lamniform Sharks from the Campanian of the Kristianstad Basin, Sweden. Palaeontology, vol. 35. pp.519-553

Siverson, M 1993 Maastrichtian Squaloid Sharks from Southern Sweden. Palaeontology, vol. 36. pp.1-18

Welton, B.J. and Farish, R.F. 1993. The Collector's Guide to Fossil Sharks and Rays from the Cretaceous of Texas 204 p.

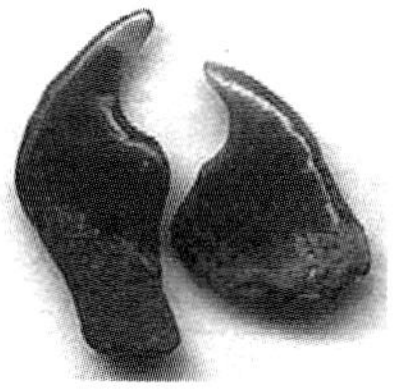

Chapter 6 — Cretaceous Microfossils

Most fossil collectors think of large vertebrate material and *Exogyra ponderosa* when talking about the North Sulphur River. However, there is a whole other world of much smaller fossils at your feet when you stand on a NSR gravel bar. For every large mosasaur vertebrae, there are dozens to hundreds of smaller fossils lying nearby. Many of these can't be found without getting on your knees and putting your nose to the ground. Other material can only be collected using bulk sampling techniques. The finds can be spectacular and the rewards are great in abundance and diversity. However, a microscope is essential to totally appreciate these fossils. So, what types of microfossils can you find on the NSR?

Dinoflagellate cysts
Acritarchs
Calcareous nannofossils
Pollens and spores (land derived)
Radiolarians
Planktonic and benthonic foraminifera (forams)
Holothurian sclerites ("sea cucumbers")
Sponge spicules
Ostracods
Scolecodonts
Shark/ray/fish/reptile microteeth
Pleistocene mammal microteeth

The strict definition of microfossils refers only to those remains which cannot be studied without the aid of a microscope. However, even larger fossils such as dinosaur bones can be studied with a microscope to investigate finer features like internal cell structures. **Micropaleontology** is the study of microfossils and is generally restricted to mineral walled microfossils such as foraminifera or fusilinids. The study of organic walled fossils is called **micropalynology**. Examples of these include pollen grains or acritarchs. Some of the fossil groups listed above can only be studied using an electron microscope because of the very small size of the fossils (100 micrometers or less). Very few fossil collectors outside of a research institution have access to this type of equipment. Most microfossils can be adequately studied with a stereoscopic microscope using 10-30X magnification. These include the forams, sclerites, spicules, ostracods, and some of the microteeth.

Dinoflagellates are single celled organisms that exhibit both plant and animal characteristics. The name means "whirling whips" and refers to the one whip-like flagellum that is used for propulsion. They have been an important element of the oceanic phytoplankton since the late Jurassic. Their fossil remains consist of the resting cyst. All fossil dinoflagellates are cysts. Very few living genera are known to encyst so this is an adaptive strategy that they have lost during the Tertiary or the fossil record is woefully underrepresented in dinoflagellates. Encysting is a response to adverse environmental conditions or following sexual reproduction in extant forms. Sudden blooms of dinoflagellates are called "red tides". They release toxic excretions which cause extensive fish kills locally.

The **acritarchs** are polymorphic groups of unicellular phytoplankton consisting of a fossilized vesicle. They range throughout the Phanerozoic but reached their apex during the early Paleozoic. Like dinoflagellates, acritarchs are of limited environmental value. Acritarch vesicles come in all shapes. They increase in diversity and abundance with increasing distance from the shoreline today.

Calcareous nannofossil remains include the plate-like scales on the cell walls and the flagella. They are derived from green algae (Division Chlorophyta) and the red algae (Division Rhodophyta). These are the two major groups of the Kingdom Plantae. Their color differences arise from different photosynthetic pigments within the cells. The green algae live predominantly in fresh water today while the red algae inhabit the world's oceans. Calcareous algae are a common constituent of carbonate rocks. Some forms of red algae are mound building. Others are coralline and were reef builders in the past and today. Calcareous nannofossils are valuable paleoenvironmental indicators. They are used to distinguish shallow water biofacies.

Pollens and **spores** are the single or multi-celled remains of plant propagation. Spores are produced by bacteria, fungi, algae, protists and a wide variety of vascular land plants. Terrestrial spores are very common microfossils. They are resistant to biological attack as well as temperature, moisture, and pressure fluctuations. They are produced in vast quantities and are dispersed rapidly and widely by wind and water. Smaller forms of spores are called "miospores" and larger forms are called "megaspores". Spores reach the marine environment by being washed into streams and rivers and carried out into estuaries and deltas. During this process they are size sorted. The larger miospores and megaspores settle out in the nearshore environments while the smaller miospores are carried out into the outer shelf. Due to their extremely small size and light-weight, they can be reworked and deposited many times over in the rock record.

Pollen grains are a specific type of multi-celled spore that is designed to reach the "female" sexual structure directly for fertilization. Pollen grains can drift for hundreds of kilometers although 99% tend to settle within one kilometer from the source plant. Very little pollen grains reach the ocean by aerial dispersal. Their greatest stratigraphic potential comes in correlating continental deposits within fluvial, lacustrine or deltaic environments.

Radiolarians are free floating protists with roughly spherical cells and thread-like psuedopodia extending radially over a delicate endoskeleton. An **endoskeleton** is an internal supporting framework for the body mass. An example is a human's bony skeleton covered by skin. The opposite is the **exoskeleton** which is an external supporting frame like crustaceans (insects, crabs, etc.) possess. The endoskeleton is composed of solid opaline silica. Radiolarians can be a primary source of silica in the generation of chert deposits. Radiolarians are entirely marine and have been around since the Cambrian at the dawn of the Paleozoic. They were especially abundant during the Cretaceous. Modern radiolarians live just seaward from the continental slope (see Fig. 7 back in Chapter 1). They feed off the plankton brought up by upwelling deep ocean currents. Radiolarians have a worldwide distribution today and live from cold polar to warm tropical waters. Different species are depth dependent and different assemblages occur at various depths in the same water column.

Foraminifera are another order of single-celled marine protists. They can either live on the sea floor (benthonic) or swim (planktonic) among the other pelagic life forms. Their soft tissue is enclosed within a test composed of secreted organic matter (tectin), agglutinated particles or calcite/aragonite. The name comes from the openings between test chambers called the foramen. Most forams are micro-omnivores that feed on bacteria, algae, other protists and small invertebrates. They in turn are fed on by worms, crustaceans, gastropods, echinoderms, and even fish. Important groups through geologic time include the Fusilinids (Pennsylvanian-Permian of Texas), while the Discorbids, Globigerinids, and the Orbitoids are more impotant in younger rocks. Forams were abundant during the late Cretaceous in the NSR study area. In general, the chalks produce more forams per bulk sample than do the marlstones. Planktonic foraminifera are very useful in intercontinental correlation of late Cretaceous rock units worldwide. Different shapes are indicative of lifestyle (benthic versus planktonic). The benthic forams *Lenticulina* and *Frondicularia* are common microfossils from the basal Ozan Formation.

Holothurian (sea cucumber) sclerites are the individual skeletal elements within the fleshy outer layer of this echinoderm. Thousands of sclerites make up the "skeleton" of a holothurian. They come in various hook and barbed shapes and are composed of calcite. Upon death of a sea cucumber and decay of the individual, the disaggregated sclerites are rapidly scattered by wave and organic (burrowing) action. Very few complete sea cucumbers are known from the fossil record even though their individual skeletal elements can be common constituents of the microfauna. This is an example of a macrofossil that will be consistently overlooked in compiled faunal listings if the microfossils are not taken into account.

Likewise, **sponges** can be easily under represented in a faunal assemblage if their microfossil remains are not carefully searched for. Demosponges (siliceous sponges) are a lot more common as individual spicules than complete specimens. The spicules generally come in some form of various planar or 3D star patterns and are composed of opaline silica. Strung together they form a mesh network that supports the shape of the sponge. Along with the radiolarians, they are a major source for chert formation. Upon death the skeleton disintegrates and the spicules are scattered along the sea floor.

Ostracods belong to the class Ostracoda of the phylum Crustacea. Their common name is "seed shrimp". They are small laterally compressed crustaceans enclosed within a protective shell. The carapace is composed of two valves that hinge above the dorsal region of the body. The carapace is ovate to kidneybean shaped and in most species is less than 3 millimeters long. Modern ostracods live in humid forest soils, freshwater ponds, and most commonly on the sea floor. Ostracods appeared with the trilobites in the early Cambrian and survive to present day. The suborder Podocopina was very diverse during the Cretaceous and can be found in marine sediments in Texas.

Scolecodonts are the jaw apparatus of annelid worms. These animals have no other hard parts to preserve outside of the jaw elements. They too will be totally overlooked as part of a fossil community unless the microfossils are sampled and analyzed. Annelid worms are common shallow water faunal elements today and there is no reason to suspect that they were not just as common during the late Cretaceous.

Vertebrate microfossils. Refer back to Table 6 in Chapter 2. Notice that out of all the late Cretaceous – Campanian Stage vertebrate localities, the Ozan Formation from northeast Texas contains the most diverse assemblage of sharks and rays. Why is that? DPS members Bruce Welton and Roger Farish (1993) took the time to adequately sample the vertebrate microfossil assemblage for teeth. Of the 23 total genera of sharks listed, only 4 could be considered common as macrofossils. These are *Cretolamna, Scapanorhynchus, Paranamotodon* and *Squalicorax*. And only 1 of the 8 rays listed can be considered a common macrofossil. That genus would be the rostral teeth of *Ischyrhiza*.

Without screening for microfossils, the true diversity of the chondrichthyan fauna from the NSR study area would never be revealed. The basal Ozan/Gober Chalk contact at Moss Creek contains a wealth of assorted fish teeth. Small fish teeth (5 mm or less) are very common along some of the NSR gravel bars. No systematic study of the bony fish microteeth has been done. Undoubtedly this would yield a greater number of fish species than is currently recorded from the NSR. The following two figures illustrate the diversity of forams and fish microteeth from the basal Ozan Formation at the Highway 34 bridge over Moss Creek.

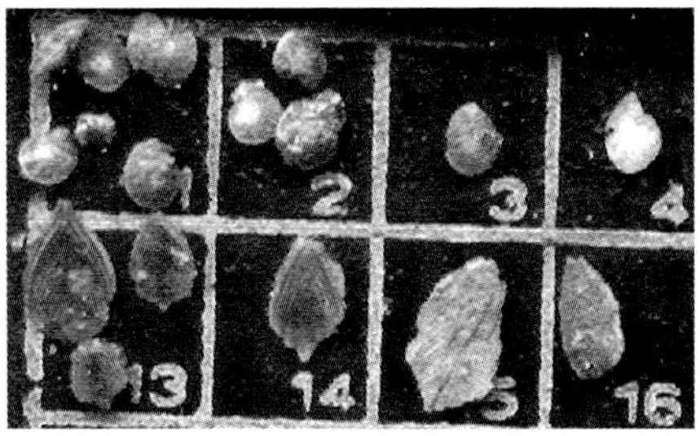

Fig. 1 Tray containing various calcareous foramanifera that have been placed onto a special microscope slide. The slide has been divided into 4mm squares. The professional biostratigrapher uses this method to do "point counts" of bulk samples to determine population densities and species diversity from sampled outcrops or cores. x3.

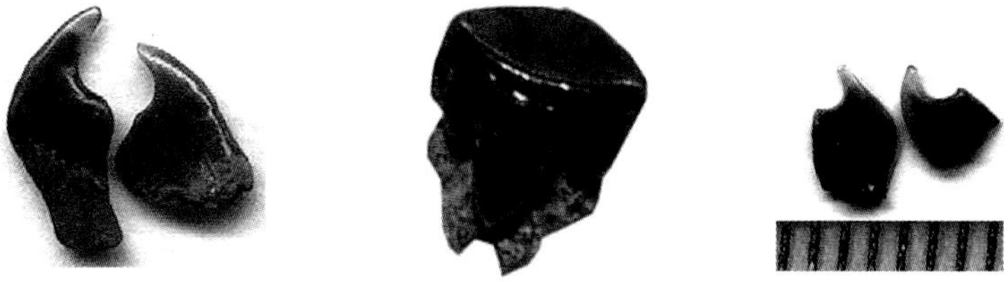

Fig. 2 Three images of assorted fish teeth. All of these fossils are only a few millimeters long in their greatest dimension and could not be easily seen without the aid of a microscope. x4, x5, x3.

Thompson et al (1991), in their comprehensive treatment on the micropaleontology of the Campanian shelf in northeast Texas, were able to distinguish the depth-zoned marine environments for all the formations encountered in the NSR. This is one of the few regional studies where different microfossils groups for the same area were analyzed, compared, and correlated with each other. Thompson studied the planktonic foraminiferans, Percival the calcareous nannofossils, Selznick the foramineral paleoecology and Heine the dinoflagellate cysts. Refer to the end of Chapter 1 of their publication for a summary of their results as it pertains to the environment of deposition for the various formations in the NSR study area. I suggest reading Thompson et al 1991 publication for a complete listing of the myriad of foraminiferans, calcarous nannofossils and dinoflagellate cysts from the Taylor Group.

Table 7. Ozan Microfossil Faunal List

AGE: Lower Mid Campanian Stage/Gulfian Series/Upper Cretaceous
FORMATION: basal Ozan Formation/lower Taylor Group
LOCATION: Just downstream from the highway 34 bridge over Moss Creek, Fannin County, northeast Texas

Foraminifera (benthic forms)
Frondicularia cf. cordata
Frondicularia cf. linearis
Lenticularia (Darbyella) sp.
Lenticularia cf. taylorensis
Frondicularia cf. goldfussi
Frondicularia sp.
Lenticularia munsteri

PHYLUM ANNELIDA
Hamulus squasmosus

PHYLUM ARTHROPODA
CLASS CRUSTACEA
ORDER DECAPODA
Misc. limb and carapace fragments

PHYLUM BRACHIOPODA
CLASS ARTICULATA
ORDER TERBRATULIDA
Two unidentified species

PHYLUM BRYOZOA
Unidentified slender, branch-like colony

PHYLUM ECHINODERMATA
CLASS ECHINOIDEA
ORDER SPATANGOIDA
Unidentified echinoid ambulacral plates and spine fragments

CLASS CRINOIDEA
ORDER ARTICULATA
Unidentified crinoid columnals

PHYLUM MOLLUSCA
CLASS GASTROPODA
Unidentified phosphatic molds

CLASS PELECYPODA
Inoceramus prisms (very abundant)
Unidentified phosphatic molds

PHYLUM CHORDATA

CLASS CHONDRICHTHYES

SHARKS

Anamotodon toddi
Carcharias aff. sanhammeri
Cretolamna appendiculata
Heterodontus sp.
Microcorax sp.
Paleogaleous navarroensis
Paranamotodon sp.
Protolamna sp
Scapanorhynus texanus
Squalicorax kaupi

Carcharias holmdelensis, C. heathi
Cantioscyllum meyeri
Ginglymostoma sp.
Microcorax crassus?
Odontaspis aculeatus

Pararhincodon groessenssi
Psuedocorax granti
Serratolamna serrata
Squatina hassei

RAYS

Ischyrhiza avonicola
Ptychotrygon triangularis
Sclerorhynchus fanninensis

Ischyrhiza mira
Rhinobatos ladoniaensis

CLASS OSTEICHTHYES

ORDER AULOPIFORMES

Enchodus ferox, E. petrosus

ORDER PYCNODONTIFORMES

Anomoeodus barbini
Unidentified pycnodont teeth

ORDER TETRAODONTIFORMES

Stephanodus sp.

References

Akers, R.E. and Akers, T.J. 1987. Texas Cretaceous Echinoids. Paleontology Section, Houston Gem and Mineral Society, 3:144 p.

Brasier, M.D. 1980. Microfossils.George Allen & Unwin, LTD. 194 p.

Case, G.R. 1982. A Pictorial Guide to Fossils. Van Nostrand Reinhold Company. 516 p.

Dane, C.H., 1929, Upper Cretaceous formations of southwestern Arkansas. Arkansas Geological Survey, 1: 215 p.

Finsley, C. 1989. A Field Guide to Fossils of Texas. Texas Monthly Press. 190 p.

Moore, R.C. (ed). 1966. Treatise on Invertebrate Paleontology (U) – Echinodermata 3(1). University of Kansas Press: 257-265.

Schafer,W. 1972. Ecology and Paleoecology of Marine Environments. University of Chicago Press. 568 p.

Shiner, H.W. and Shrock, R.R. 1944. Index Fossils of North America. John Wiley & Sons, Inc. 838 p.

Thompson, L.B., Heine, C.J., Percival, S.F.,Jr. and Selznick, M.R. 1991. Stratigraphy and micropaleontology of the Campanian shelf in northeast Texas. Micropaleontology Special Publication, 5:148 p.

Welton, B.J. and Farish, R.F. 1993. The collector's guide to fossil sharks and rays from the Cretaceous of Texas. BeforeTIME Publishing. 204 p.

The North Sulphur River drainage as it appears today has been sculpted by geologic processes that have occurred during the last 12,000 years. This includes the end of the last great continental glacial advance and retreat to the present day. The youngest major chronostratigraphic subdivision of geologic time is the Quaternary System. It encompasses the four major glaciations of the late Cenozoic including the last great Ice Age. The present day topography of the NSR study area was firmly established by the end of the Pleistocene 10,000 years ago.

The classical subdivision of the North American Quaternary System involves four glaciation and three interglacial events. In order from youngest to oldest they are;

Wisconsinan Glaciation
Sangomanian Interglacial
Illinoian Glaciation
Yarmouth Interglacial
Kansan Glaciation
Aftonian Interglacial
Nebraskan Glaciation

The glacial events were named after localities in the United States where glacial deposits are well developed and the relative ages are firmly established. Research in the last few decades has demonstrated that the two older glaciations (Kansan and Nebraskan) are actually a complex series of time transgressive events and not two distinct episodes of glaciation. These terms have been thrown out in favor of the terminology pre-Illinoian A, B, C, D, etc. The oldest glaciation commenced in the Miocene approximately 6.0 million years ago in the northern latitudes of Alaska. The last Ice age (the Wisconsinan) ended roughly 10,000 years ago. Of course, the Ice Age ended at different times at different latitudes as the continental ice sheets retreated towards the poles.

We are technically still in the "Ice Age". There is no hard scientific data to predict if the Earth is in a prolonged period of warming or if the Ice Ages will commence again in another 10,000 years. Concerns over global warming caused by humans polluting the atmosphere and tearing a hole in the ozone layer of the outer atmosphere could be premature. This phenomenon may be part of a natural cycle on a time scale much longer than man's current ability to measure. Man may merely be accelerating a process that is occurring naturally anyways.

There were two main epicenters for glaciation on the North American continent as shown in Fig. 1. The eastern one is termed the **Laurentide Ice Sheet** and was centered on the Canadian Shield north of the United States border. It originated west of the Hudson Bay area as the Keewatin Ice Center. It has been estimated that the maximum thickness of the Laurentide ice sheet was 3000 meters (almost 10,000 feet) thick. It is responsible for the formation of the Great Lakes in the upper Midwest. They were formed when the ice margin retreated around 14,000 years ago. As the ice sheet advanced, it scraped and gouged the bedrock and soil cover in front of it. This material was scooped up and carried along like a conveyor belt within the migrating ice sheet. When climatic conditions changed and the ice sheet stabilized, it would drop its sediment load in glacial deposits called terminal moraines. These linear features can be a hundred feet tall and extend for miles. The ice sheet would eventually retreat and melt. These moraines would act like natural dikes and levees to impound the meltwater behind them. Huge Pleistocene lakes were formed along the margins of the ice sheet during the Late Wisconsinan. The last major ice advance in the Great Lakes region covered the Lake Superior Basin and is dated to 10,000 years ago.

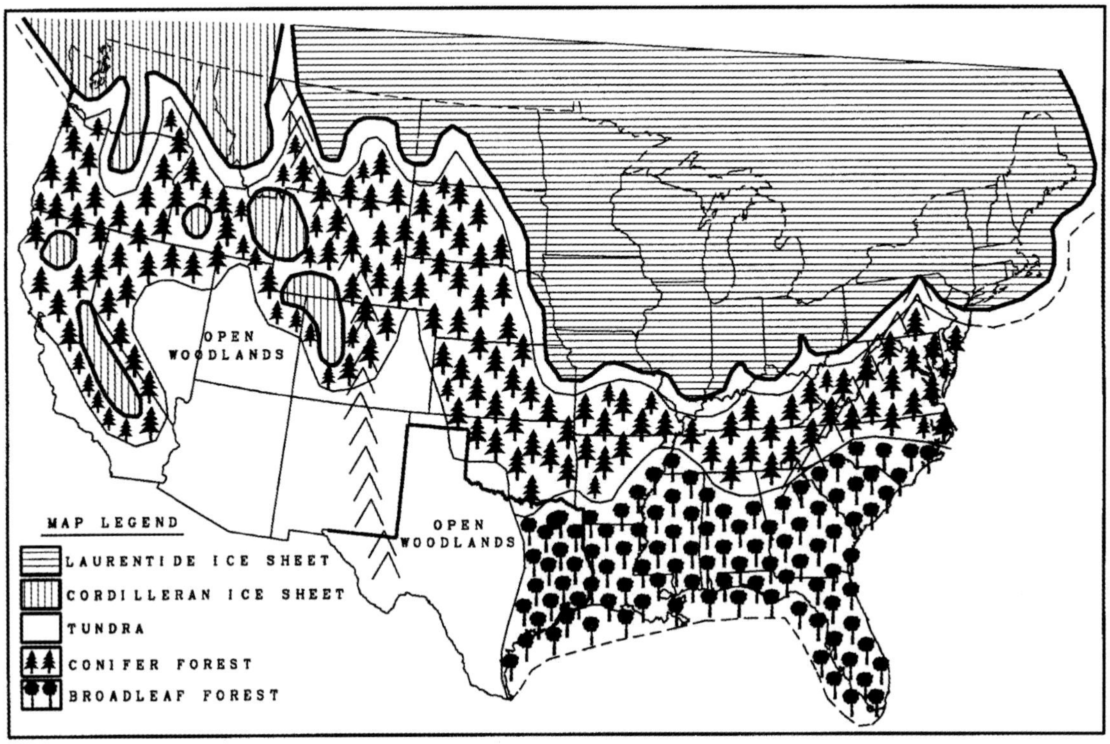

Fig. 1 Generalized vegetational model of North America during the last Glacial Maximum, 18-20K years ago. Modified after Canby (1979), McDonald (1981) and McDonald (1984).

Post glacial sea level rise ended about 7000 years ago. During the glacial maxima, huge volumes of seawater were locked up in the continental ice sheets worldwide. Present day sea level was as much as 300 feet lower during the Ice Age. PaleoIndians could have easily walked from Miami to the tip of the Florida Keys without ever getting their moccasins wet. Cuba would have been a short canoe trip away. The Hudson Bay lowlands of Canada did not deglaciate until 6500 years ago. The further north you go, the later the glaciation ended. No expert can pick a specific date for the end of the Pleistocene glaciation as it varied based on elevation, topography, and distance away from the equator.

In the mountains of western North America, the **Cordilleran Ice Sheet** formed by the coalescence of large piedmont glaciers in the Cascade and Rocky Mountain Ranges of Canada. This solid ice sheet migrated as far south as the middle of Idaho. The higher mountain ranges to the south all had alpine glaciers at the higher elevations. Try to imagine the Front Range of Colorado covered in a sheet of ice with just the tops of the mountain peaks jutting out. They would have resembled the modern day Himalayan Mountains of Nepal or the Southern Alps of New Zealand.

No glaciation ever reached as far south as Texas (see Fig. 1). The Laurentide ice sheet extended as far south as NE Kansas during the last glacial maximum approximately 18,000 years ago. The edge of the continental ice sheet was not a towering cliff but a gently tapering slope like small alpine glaciers found today. In front of the advancing ice sheet would have been a tundra "buffer" zone seperating the ice from the dense conifer forests that extended from one coastline to the other. The remnants of this magnificent forest type composed of pines and other evergreen trees can still be seen in the Pacific Northwest today. South of this zone the forest composition changed to broadleaf trees such as oaks, maples, birches and many others. When the ice age ended and the ice sheets started retreating northwards; the forests migrated with it.

There are no direct glacial deposits such as tills, loess, or moraines in northeast Texas. However, the late Wisconsinan Ice Age did influence the global climate and locally controlled the type and rate of sedimentation that took place along the NSR drainage basin. Fig. 2 is the local Quaternary stratigraphic column for the NSR study area *prior to channelization* of the NSR in 1929.

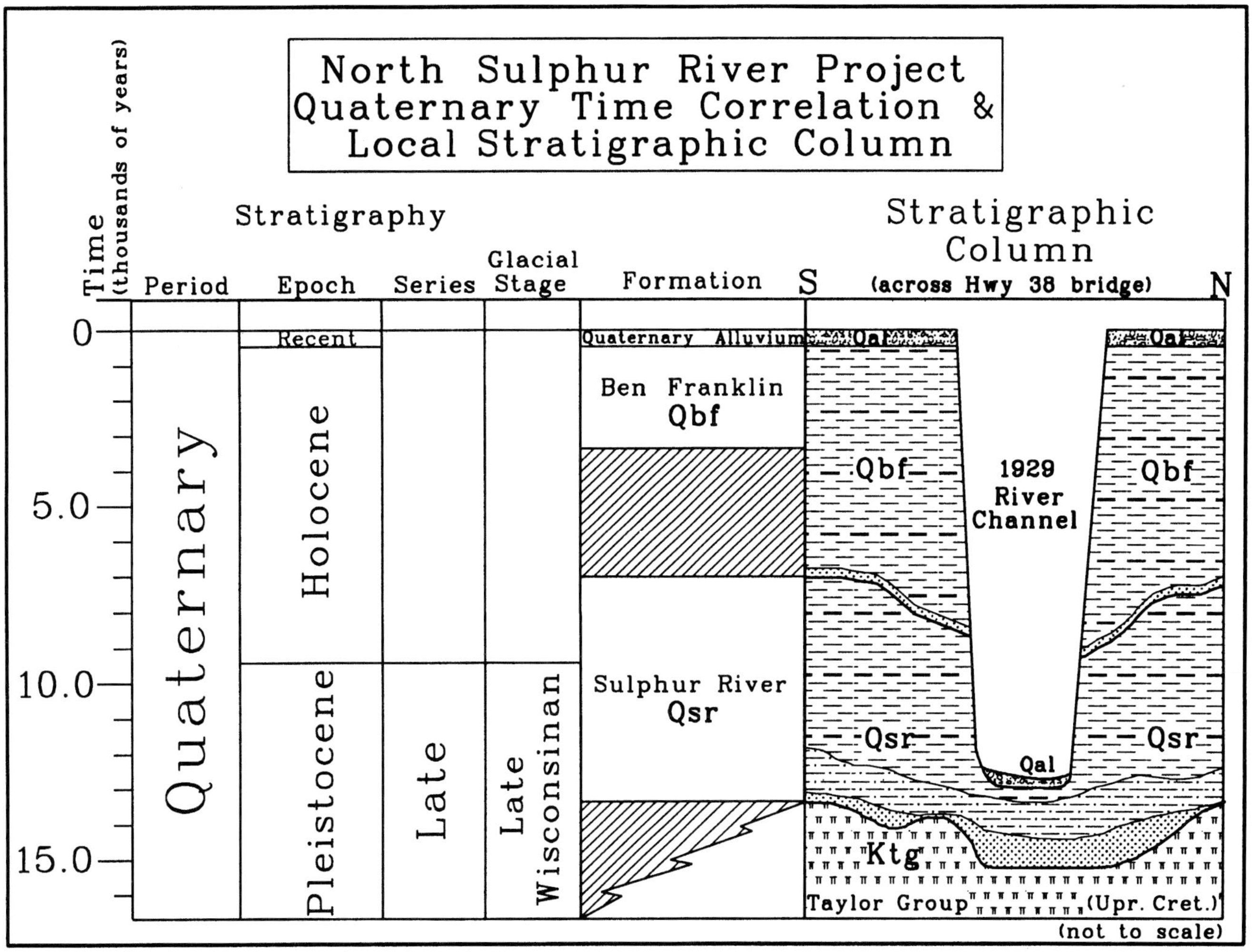

Fig. 2. Local Quaternary Stratigraphic Column for the North Sulphur River.

At the end of the Cretaceous 66 million years ago the great Western Interior Sea dried up. The Gulf of Mexico retreated to its present day location during the lower Tertiary. Prior to the late Pleistocene, the NSR valley experienced an interval of unknown length in which the late Cretaceous Taylor Group was eroded into a series of hillocks and divides. Changing climatic conditions caused the Pleistocene NSR to begin a period of net aggradation. This resulted in the formation of a floodplain and deposition of the lower Sulphur River Formation (Qsr) shown in Fig. 2. The basal unit of the lower Sulphur River Formation consists of iron-stained, fine pebble to coarse sand gravel. It grades upward into a gray sandy silt. This fining upward sequence of sedimentation is indicative of a meandering stream. This is also the first chance that Cretaceous fossils had to be eroded out of the Taylor Group and deposited in younger (late Pleistocene) sediments.

There are numerous periods of channel downcutting into the floodplain recorded within the Sulphur River Formation. This is easy to imagine if you have ever studied a stream or river after a major flood. The gravel bars (depositional features) and the slumping cut banks (erosional features) do not remain static but migrate a little bit after each flooding event. Over the years this changes the course of the river within the confines of the floodplain. Carcasses of Pleistocene mammals washed into the river and would become lodged with other debris on a gravel bar during a heavy rain. With any luck they would be buried in subsequent flooding events.

In 1963, Slaughter and Hoover reported on the results of quarrying efforts for micro vertebrates in the Sulphur River Formation. Several tons of matrix were removed from five quarry sites within 600 yards of the Hwy 38 bridge north of Ben Franklin. The samples were washed and screened for micro vertebrate remains. The results are tabulated in Table 8 – "Ben Franklin" local vertebrate fauna. At quarry number 5 charcoal was recovered from a prehistoric burn pit in the upper Sulphur River Formation. The hearth was approximately seven feet in diameter and was at the western edge of a Pleistocene pond. Radiocarbon dating of the charcoal yielded a date of 9550 +/- 375 years before present (BP). Hearth charcoal is an excellent dating material because it is resistant to chemical attack and as an undisturbed feature precludes redeposition at a later date. Within fifty feet of the hearth an antler artifact was recovered *in situ*. The antler artifact cannot be accurately dated in relation to the burn pit but the two elements together indicate that paleoIndians visited the NSR valley during the late Pleistocene.

The climate of northeast Texas 9500 years ago can be surmised by plotting extant mammals current ranges in North America today versus their fossil ranges during the Pleistocene. The abundance of eastern and woodland species indicates that cooler summer temperatures prevailed then and that the overall moisture was probably greater than it is today. This could mean either greater total rainfall amounts or less evaporation during the cooler summer months. Present day Delta County has a mean annual average temperature of 64 degrees Faranheit and the average annual rainfall is 40 inches. In eastern North America, the climate warming following the glacial retreat reached its maximum around 6000 BP. It has been slowly cooling off since then. The climate was also much wetter than it is today. For example, in the Chicago area it has been calculated that the annual precipitation was 8 inches greater than it is today. In summary, the climate in northeast Texas at the end of last Ice Age involved milder, wetter summers than today. In addition, the winters were actually less severe than today.

Deposition of the Sulphur River Formation continued until approximately 7000 years ago when a period of nondeposition ensued. A layer of soil developed on top of the Sulphur River Formation floodplain. This was followed by renewed downcutting of the floodplain around 2800 BP. Valley filling continued during the Holocene with deposition of the Ben Franklin Formation (Qbf). This was the second episode in which late Cretaceous fossils could be reworked into younger (Holocene) deposits by fluvial processes. The distinctive late Pleistocene mammalian fauna was totally replaced by a "modern" one everwhere in the lower 48 states by 7000 years ago at the very latest.

Prior to 1929, the NSR valley and floodplain was approximately half a mile wide and consisted of a lazy meandering stream in a swampy lowland. The bottomlands were prone to flooding and darn near impossible to farm. In 1929 the Texas Reclamation Department channelized the North and South Sulphur Rivers. This halved the distance traversed by the NSR and consequently steepened the gradient of the stream. The river channel was only 12 feet deep and 30 feet wide prior to 1929. As of 1974 it was 30 feet deep and 200 wide. Before 1929 all 38 tributaries fed in from the north. Since channelization, 45 new tributaries have developed on the south side of the straightened river. Over 100 new headcuts enter the river from the north. This is the third episode of potential erosion and reworking of Cretaceous, Pleistocene, and Holocene fossils in the NSR study area.

All the gravel bars (or Quaternary Alluvium (Qal) of Fig. 2) that we collect along the channelized extent of the NSR today have been deposited in the last 70 years. Prior to the 1929 channelization, the NSR did not cut down into the late Cretaceous deposits in any appreciable amount. What has been good for the farmer is also good for the collector. And what is now bad for the farmer (increased headcut erosion, loss of farmland) is also good for the collector as it continually increases the amount of fossiliferous bedrock exposed. The proposed damming of the NSR valley to create another water reservoir for the greater DFW area may effectively end fossil collecting as we know it. Hopefully this won't happen in our lifetime!

Doering, J.A. 1963. Correlation of the Sulphur River Formation. Journal of the Graduate Research Center, 36(3):149-157.

Ehlers, J. 1996. Quaternary and Glacial Geology. John Wiley & Sons. 578 p.

Rainey, M. 1974. The Quaternary stratigraphy of the North Sulphur River. Southern Methodist University unpublished Master's thesis, ? p.

Slaughter, B.H. and Hoover, B.R. 1963. Sulphur River Formation and the Pleistocene mammals of the Ben Franklin local fauna. Journal of the Graduate Research Center, 36:132-148.

Slaughter, B.H. and Ritchie, R. 1963. Pleistocene mammals of the Clear Creek local fauna, Denton County, Texas. Journal of the Graduate Research Center, 31:117-131.

Slaughter, B.H. 1964. Geological survey and appraisal of the paleontological resources of the Cooper Reservoir Basin, Delta and Hopkins Counties, Texas. Fondren Sciences Service, 6:11 p.

Stearn, Colin W., Carroll, Robert, L. and Clark, Thomas H. (editors), 1979. Geological Evolution of North America. John Wiley & Sons. 566 p.

Wright, H.E., Jr. and Frey, D.G. 1965. The Quaternary of the United States. Princeton University Press. 922 p.

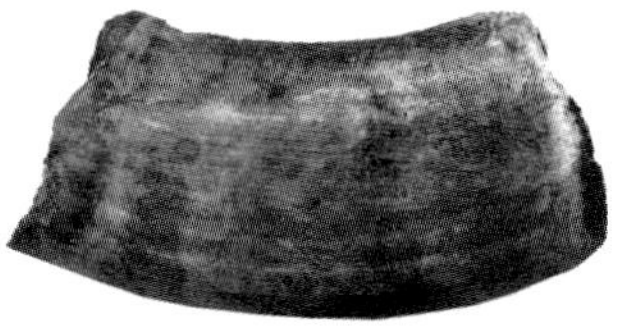

Chapter 8 Pleistocene Paleontology

Table 8 summarizes the distribution of macro vertebrates, including those brought to the NSR symposium. These fossils are dominantly individual bones from larger mammals that were collected by walking the gravel bars (Location B). Other late Pleistocene fossils from the NSR study area include the invertebrate freshwater clams and gastropods. A majority of the fossils listed are **extant** (living) species. However, some of the larger mammals are now extinct and represent species that died out at the end of the last glacial period. Please keep in mind that the late Pleistocene fossils recovered from the gravel bars range in age from approximately 12,500 years ago up to modern times. That is why you have a mix of more typical "ice age" mammals along with contemporary forms.

I will briefly review the life habits, relationships, presumed distribution and estimated extinction of some of the larger mammalian forms found in the NSR. But first let us take a look at the one large reptile whose remains were brought to the symposium. That would be the giant land tortoise *Geochelone*. Late Pleistocene land tortoises are found on every continent excluding Antarctica and Australia. Today, their range is restricted to remote oceanic islands such as the Galapagos. It is not a loss of habitat that has greatly reduced tortoises in aerial extent and numbers. It is the presence of man! Much like the large flightless birds (such as the moas of New Zealand), primitive man hunted them to extinction wherever and whenever the two came into contact. Slow moving land tortoises were easy prey for paleoIndians. It is generally assumed that giant land tortoises indicate a very warm temperate to subtropical climate with little chance of a winter freeze.

The most famous and easily identifiable remains of late Pleistocene mammals are the mammoths and mastodons. Tusk fragments and molars are the common fossil elements. Both are members of the order Proboscidea that includes the modern day African elephant *Loxodonta africana*. African elephants reach heights of 3.5 meters and weigh up to 7,500 kilograms (or 8.3 tons). The Columbian mammoth *Elephas* (=*Mammuthus*) *columbia* was of comparable size with the mature adults reaching 3.6 meters in height. Mammoths were extremely common in the Pleistocene of North America west of the Mississippi River. Slaughter (1960) reported 15 semiarticulated skeletons of *Elephas columbia* from the Moore Pit local fauna along the Trinity River in south Dallas, Texas. These deposits range between 50,000-37,000 years BP during the Sangomanian Interglacial period. This was prior to the last Wisconsinan Glaciation. Slaughter estimated that <u>not</u> one in one thousand mammoth bones are recovered in commercial sand and gravel excavations.

Mammoths preferred open plains and prairies to forests. They invaded North America approximately 1.5 million years ago and were hunted by paleolithic man as late as 11,000 years ago. The dwarf mammoth, *Elephas exilis* found on Santa Rosa Island, offshore California, stood two meters high at the shoulder and survived until 5500 years ago. In the last ten years, dwarf mammoth remains dating to 3750 years ago have been recovered from a remote island in the Aluetian Island Chain between Russia and Alaska. To put this in historical perspective, at the same time that ancient Britons were constructing Stonehenge, dwarf mammoths were roaming the tundra on the other side of the globe.

In both cases, the lack of human occupation on the island probably contributed to the dwarf mammoth surviving later then its larger relatives on the mainland. The initial breeding populations were probably large size mammoths that were trapped on the islands as the glaciers melted and sea level rose.

An ecological dilema ensued. The island presented a limited and dwindling food resource. Overgrazing would put the entire population at risk. The mammoths responded by becoming progressively smaller with each generation. This resulted in a smaller total food intake per individual. This is an example of adaptive evolution at work.

Mastodons are considerably smaller than their mammoth cousins. The American mastodon, *Mammut americanus* stood over two to three meters high at the shoulders. Judging from its dentition and its Pleistocene distribution, mastodons were browsing animals that preferred coniferous forest habitat. The only continent with mastodons during the Pleistocene is North America. They survived as late as 8000 years ago. There are no kill sites linking humans and mastodons together. However, it seems reasonable to assume that they were hunted by early Native American ancestors.

Isolated horse molars are very common fossils from the NSR (plate 19, Figs. 4 and 5). Unfortunately, it is hard to distinguish a four year old from a 400 year old versus a 4,000 year old molar in the field. Prehistoric horses have a long and varied history during the Tertiary of North America. One of Charles Darwin's main lines of evidence in support of his theory of evolution is horse lineages from the fossil record. The earliest horses were small (the size of a medium dog), and had five toes. Horses grew larger and reduced the number of toes throughout the Cenozoic as they became "better" adapted to life on an open savanna.

There are eight living species of *Equus* in the world today and numerous breeds that are biologically capable of successfully reproducing with each other. There are 12 described fossil horse species from the Pleistocene of Texas alone! There has been an over splitting of fossil horses by American paleontologists in the past and some of them are not valid. That is one reason why it is difficult to identify a horse molar to the species level.

Equus became extinct in the Americas by 8500 years BP. It is hard to explain their demise on this continent as there is no identifiable ecological mechanism to have triggered the enormous loss of habitat required to cause its extinction. The modern horse was domesticated from the tarpan of eastern Europe and western Asia. The horse was reintroduced in North America by the Spaniards in the late 1500's/early 1600's. From a very small population that escaped in Mexico, horses rapidly spread northward and proliferated along the length of the Great Plains. By the time of the Louisiana Purchase in 1804, numerous Indian tribes had already shifted from a fixed village-based existence to a nomadic lifestyle based on the horse. The expansive prairies of the Great Plains, which had been a barrier to migration, were suddenly open to exploitation by the Indians. What is even more phenomenal is that this all happened in the space of a few generations.

The American Plains Indian's principal food source was the buffalo *Bison bison*. In the early post glacial period, North American paleoIndians hunted *Bison antiquus* and *Bison occidentalis*. Skull, horns and leg bones of the modern buffalo *Bison bison* have been collected from the NSR. The three species of bison cannot be distinguished from one another except by the size of their horn cores. The horns grew progressively smaller from the older to the younger lineages. In the early 1800's, bison grazed in huge herds that literally covered the prairie as far as the eye could see. It is estimated that the number of buffalo on the Great Plains prior to 1850 reached 50 million individuals. There were two great herds; one centered in Texas and Oklahoma and the other further north in the plain states bordering Canada. In terms of sheer biomass, the Great Plains of the early 1800's rivaled the Plains of Serengeti in Africa. By 1906 only a few hundred buffalo remained, the rest wiped out by the westward expansion of American civilization.

Other large extinct herbivores from the NSR Pleistocene include llamas, camels, ground sloths, tapirs, long-nosed peccaries and the woodland muskox. There are three ground sloths that potentially inhabited northeast Texas in the post-glacial Pleistocene. They are *Megalonyx*, *Eremotherium* (=*Megatherium*) and *Paramylodon*. Jefferson's ground sloth *Megalonyx jeffersoni* (named after our illustrious president Jefferson) is a woodland dweller that migrated upward from South America in the Pliocene. It persisted until the late Pleistocene approximately 9400 years ago. *Eremotherium* was a giant ground sloth that reached a height over 5 meters (17 feet) when standing on its rear haunches. Ecologically it filled the niche occupied by giraffes in Africa. Isolated incisors and claws of ground sloths (Plate 18, Figs. 7, 8, and 12) can be found in the gravel bars along the NSR. The third type of ground sloth is *Paramylodon.* It stood about four feet tall over its back. There are no known kill sites with ground sloth bones but it seems that they would have been easy prey for paleoIndians if they were not already extinct.

Glyptodonts and the giant armadillo were already extinct by the post glacial late Pleistocene in North America. They are common in Texas from Sangomanian age deposits (prior to the last Wisconsinian glaciation). We do find isolated armor scutes and tail sections (Plate 18, Figs. 2 and 3) from the giant armadillo *Dasypus bellus* in the NSR. This species was roughly three times the size of a modern-day armadillo so common as road kill in Texas today. The youngest reported occurrence anywhere in the United States is from the Ben Franklin local fauna by Slaughter (1963). The pampathere *Holmesia septentrionalis* is another armadillo-looking creature that migrated up from South America. The youngest dated occurrence of it is 9880 BP from Hornsby Springs, Florida by Kurten and Anderson (1980). Isolated scutes from pampatheres have been recovered from the NSR gravel bars (Plate 19, Figs. 10 and 11).

Camelops (=*Escatius*) is the North American equivalent to the modern-day dromedaries of southwest Asia and north Africa. It stood taller as its legs were 20% longer than today's counterpart. Skeletal morphology indicates a single mid-dorsal hump. NSR fossil specimens (Plate 19, Figs. 6-8) indicate a post glacial extinction of Pleistocene camels in North America. Once again, there is no valid reason for the demise of this genus on this continent. Like the horse, camels evolved in North America. The U.S. Calvary reintroduced camels to the American Southwest in the second half of the last century. They thrived on native southwestern shrubs such as mesquite and creosote. There are no known archeological kill sites so extermination by man is not probable.

Tapirs are strange looking pig-like animals that are more closely related to horses and rhinoceroses. They have a hairless body and are characterized by a short, elephantine-looking snout. There are four living species, three of them in the tropical Americas. There are four extinct species from North America. The NSR remains (Plate 19, Fig. 3) are some of the youngest recovered in the country. *Mylohyus* is the "long-nosed" peccary of the North American Pleistocene. It was very common in Texas and is regarded as a woodland browser and omnivore. Once again, the NSR remains are some of the youngest recovered in the country. Modern-day peccaries prefer warm-temperate to tropical environments but evidently endured cooler temperatures during the late Pleistocene.

Musk ox remains include a beautiful skull collected in 1991 by Lewis Smith and Harold Finley from the NSR. The specimen is now on display at the Fort Worth Museum of Natural History. The Woodland musk ox, *Symbos cavifrons*, is strictly a Pleistocene North American form. It was taller and more slender than the extant musk ox and were grazers inhabiting woodlands and surrounding open plains. The previously youngest terminal date for its extinction is 11,000 years ago based on remains from Kalamazoo, Michigan. The skull from the NSR may be younger but since it was collected so long ago there is no way to verify its age by collecting soil samples from the spot it was collected at.

The order Rodentia (rodents) is a group with large size variation. It includes very small forms such as the pine vole, gray squirrel and the cotton rat. It also includes larger forms such as the eastern pocket gopher, the muskrat and the giant beaver *Casteroides ohioensis*. The highly modified incisors (Plate 18, Fig. 4), used to gnaw down trees and saplings at their base, are commonly preserved as fossils in the late Pleistocene deposits of Texas. A mature adult reached lengths of four meters (13 feet) from head to tail. They went extinct at the end of Wisconinan glaciation.

So far we have only discussed herbivores, insectivores, and omnivores. There were also plenty of carnivores around to prey upon the young and the weak. These included the extant black bear *Ursus americanus* (Plate 18, Fig. 5), the modern coyote *Canis latrans* and an unidentified feline from the Ben Franklin local fauna. There is no reason why there would not have also been wolves and pumas on the prowl along the NSR during the late Pleistocene.

If we could transport ourselves back in time 10,000 years to the late Pleistocene along the NSR valley, we would see a world not much different from our own. The overall climate would have been the same. The temperature extremes would have been less varied. We would not have experienced the intense summer heat for months on end. There was even less frequency of "blue northerners" during the winter months. Overnight freezes would have been rare. Overall moisture content would have been up and the area would have been considered woodlands. The open grasslands of the prairies, if existent at all, would have been much further to the west than they are present day. The last Wisconsinan ice sheet, which never reached further south than Kansas, was well on its way retreating back into Canada. There would be no obvious indications that the ice age glaciation had ever affected this area.

However, one big difference would have jumped out at us as we surveyed the large mammalian herbivores browsing, grazing, and munching their way up and down the length of the valley. Animal groups only seen in Africa today would have been commonplace then. Large quadrapeds included elephants, camels, llamas, horses, sloths, peccaries, bison, antelope, musk ox and deer. All would have been foraging for food in the open breaks and along the forest floor. And even the recognizable modern forms would be of gargantuan size. Imagine giant armadillos and beavers larger than an adult human. In addition, a full complement of recognizable small mammals, birds, reptiles and amphibians dwelt in the area.

What caused the extinction of these large mammals at the end of the last ice age? How come the majority of the large mammalian herbivores survived four major episodes of glaciation (and three inter-glacial periods) only to become extinct at the end of the last one? Why did it not affect the smaller mammals such as the rodents? Why were the large marine mammals (such as the whales) not affected?

There is no broad consensus among Pleistocene reasearchers even today. The two main camps involve climatic reasons or cultural (hunting) causes. In reality it is probably a complex relationship with both factors contributing to the demise of large land mammals on the North American continent.

Martin (1967) gives a very comprehensive survey of North American Pleistocene megafaunal extinctions. There were fifty large genera of land based mammals in the Pleistocene of the United States. Four became extinct at the end of the Kansan, three at the end of the Yarmouth interglacial, none at the end of the Illinoian and 33 at the end of the Wisconsin. In each case but the last one, the megafauna was replaced (niche replacement) by new species. Ecologists refer to niche as what an organism does in its natural environment. Habitat is where it lives. The major difference between the earlier Pleistocene extinctions and the last one is the presence of man. Scientists still do not have one definative answer regarding the late Pleistocene megafaunal extinctions. A drastic change in climate, coupled with food hunting by paleoIndians both contributed to their demise.

Fig. 1 is a graphic representation of some of the larger mammalian faunal elements found along the NSR valley during the late Pleistocene epoch approximately 9800 years ago. The climate would have been much like present day except without the severe hot and cold temperature extremes. The receding continental glaciers would have been situated hundreds of miles to the north and had never directly shaped the landscape of northeast Texas. The NSR valley itself would have been heavily wooded with the surrounding divides being open savannas.

In the scene depicted above, two charging mastodons, *Mammut americanus*, trumpet their arrival as they enter the NSR floodplain from the terrace above. A paleo-camel, *Camelops sp.*, looks on with mild interest. The giant beaver *Casteroides ohioensis* also ponders the elephant's noisy approach. In the background, two different giant sloths forage in the tree canopy foilage. To the left is *Megalonyx jeffersoni* and to the right is *Eremotherium sp*. Seemingly oblivious to it all is the tapir *Tapirus sp*. A small flock of seagulls circles overhead. They are attracted to the marshes and wetlands along the valley floor caused by the numerous beaver dams along the waterway.

Table 8. – Pleistocene Vertebrate Fossils from the NSR Study Area:

Location A -"Ben Franklin" Local Vertebrate Fauna (1963).
Age: Late Pleistocene – Late Wisconsinan Glacial Stage (approx. 11,000 - 9500 years ago)
Formation: Ben Franklin Formation
Exact Location: 5 quarries within 600 yards from the Hwy 38 bridge north of Ben Franklin, Delta County, Texas.

Location B – Gravel Bars along the length of the NSR Study Area channel bottom (1999).
Age: Late Pleistocene to Recent (12,000 – 500 years ago)
Formation: Ben Franklin and Sulphur River Formations

Class	Order	Scientific Name	Common Name	Location
Amphibia				
	Urodela	*Ambystoma tigrinum*	Salamander	A
	Anura	*Rana cf. catesbeiana*	Bull Frog	A
Reptilia				
	Chelonia	*Geochelone sp.*	Land Tortoise	B
	Crocodilia	*Alligator mississippiensis*	Alligator	B
	Squamata	*Thamnophis sp.*	Ribbon Snake	A
		Coluber (or Masticophis)	Racer Snake	A
Mammalia				
	Insectivora	*Sorex cinereus*	Long-tailed Shrew	A
		Blarina sp.	Short-tailed Shrew	A
	Edentata	*Dasypus bellus*	Armadillo	A, B
		Holmesina septentrionalis	Pampathere	B
	Rodentia	*Casteroides ohioensis*	Giant Beaver	A, B
		Geomys sp.	Eastern pocket Gopher	A
		Microtus pennsylvanicus	Vole	A
		Microtus pinetorum	Pine Vole	A
		Neotoma sp.	Wood Rat	A
		Ondatra zibethicus	Muskrat	A
		? Reithrodontomys sp.	Harvest Mouse	A
		Sigmadon hispidus	Cotton Rat	A
		Spermophilus franklini	Squirrel	A
		Synaptomys cf. cooperi	Lemming	A
	Carnivora	*Canis latrans*	Coyote	A
		Ursus americanus	Black Bear	B
		?Felix sp.	Leopard	A
	Proboscidea	*Elephas cf. columbia*	Columbian Elephant	A, B
		Mammut americanus	American Mastadon	A, B
	Lagomorpha	Rabbit	Rabbit	A
	Xenarthra	*? Eremotherium sp.*	Ground Sloth	B
		Megalonyx jeffersoni	Ground Sloth	B
	Artiodactyla	*Bison bison*	Modern Buffalo	A, B
		? Antilocapra americana	Pronghorn	A
		Camelops sp.	Camel	B
		Llama	Llama	B
		Mylohyus sp.	Long-Nosed Peccary	A, B
		Odocoilus virginianus	White-tail Deer	A, B
		Symbos cavifrons	Woodland Musk Ox	B
	Perissodactyla	*Equus sp.*	Modern Horse	A, B
		Tapirus sp.	Tapir	B

Doering, J.A. 1963. Correlation of the Sulphur River Formation. Journal of the Graduate Research Center, 36(3):149-157.

Martin, P.S. and Wright, H.E., Jr. 1967. Pleistocene Extinctions–Search for a Cause. Yale University Press. 453 p.

Rainey, M. 1974. The Quaternary stratigraphy of the North Sulphur River. Southern Methodist University unpublished Master's thesis, ? p.

Slaughter, B.H. and Hoover, B.R. 1963. Sulphur River Formation and the Pleistocene mammals of the Ben Franklin local fauna. Journal of the Graduate Research Center, 36:132-148.

Slaughter, B.H. and Ritchie, R. 1963. Pleistocene mammals of the Clear Creek local fauna, Denton County, Texas. Journal of the Graduate Research Center, 31:117-131.

Slaughter, B.H. 1964. Geological survey and appraisal of the paleontological resources of the Cooper Reservoir Basin, Delta and Hopkins Counties, Texas. Fondren Sciences Service, 6:11 p.

Slaughter, B.H. 1966. The Moore Pit Local Fauna; Pleistocene of Texas. Journal of Paleontology. 40(1): 78-91.

Wright, H.E., Jr. and Frey, D.G. 1965. The Quaternary of the United States. Princeton University Press. 922 p.

One of the greatest advantages to collecting fossils along the North Sulphur River versus other parts of Texas is the easy access to the river. There are numerous bridges and low water crossings between the Highway 68 (Wolfe City - Bonham) to the west and the Highway 24 (Paris - Commerce) bridges to the east. Please refer to Fig. 1 as it highlights some of the more common access points to the river. Just park as close as possible to the bridge off the shoulder of the road, and hike down to the river bottom. The NSR is generally a very steep sided channel with riverbanks in excess of thirty feet tall. Do not assume that you will be able to climb out of it from any point in between road bridges. Unless you take two vehicles and park one at another bridge beforehand, you will have to leave the river at the same point you entered it.

Always wear clothing you don't mind getting very muddy and possibly very wet. It is always a good idea to bring an extra set of pants and shoes to change into when you get back to your vehicle at the end of the day. Most people collect by walking upstream or downstream from their take in point, searching the gravel bars and channel walls for fossils. You generally cross and recross the river numerous times as you move from one gravelbar to another. Luckily, the river is normally one to two feet deep so that a river crossing is not a dangerous maneuver. The vertebrate material, petrified wood, and large *Exogyra ponderosa* oysters can be seen while standing up. The smaller pieces such as the shark's teeth and gastropods etc. will be collected by crawling on your knees once you find a sweet spot containing them.

A majority of the fossils collected from the NSR will be loose material retrieved from the gravel bars and riverbanks. You generally do not need any special digging tools such as a shovel or spade. A sturdy backpack, numerous plastic baggies and film/pill canisters, a screwdriver or small chisel, and a rock hammer is all the collecting gear you need. Some collectors will screen certain areas for microteeth. In that case, you will need a set of screens or sieves and a hand trowel to dig up sand and gravel samples. Luckly, the river is there to supply the water needed to wash the samples thru the sieves. It is always prudent to carry along a small notebook to take notes in case something exceptional is found.

If you are fortunate enough to find an articulated specimen of some Cretaceous or Pleistocene vertebrate, then bigger collecting tools will be required. But first off, consider contacting the DPS or the DMNH for assistance. Your find may be of scientific importance. For the excavation itself you will need shovels. You will also need five gallon buckets to haul away material in. First, collect all the loose vertebrate material that is lying on the river or creek bottom. Next, carefully examine the riverbank or creek bottom for any exposed fossils. If you find some pieces sticking out of the matrix, take a deep breath and remain calm. Your first impulse is to haphazardly start digging out the fossils. No! No! No! Step back and try to ascertain the position of the pieces to see if it is an articulated specimen. If it is an association, pull out the notebook and sketch out the position of the fossils. Trust me, once you start excavating the specimen, it will be a whole lot easier to reassemble it at home if you take the time to record the fossil locations during the removal process. It is not a bad idea to have a camera ready while you extract it from the matrix. The pictures can be used later for reference.

If at any time the task seems overwhelming, or you are not confident about removing it, or you think it is a real significant find – please leave the specimen alone and seek help. You can contact any of the Society's officers and we will send the FBI (Fossil Bureau of Investigation) swat team to investigate. They can make a recommendation as to the value of the specimen and how to proceed. Or you can contact the geology department at Southern Methodist University or the Dallas Museum of Natural History. Both of these institutions are valuable sources of assistance. The Dallas Paleontological Society is a non-profit organization promoting the science of paleontology. The Society is not interested in helping people who want to sell their vertebrate finds for commercial interests.

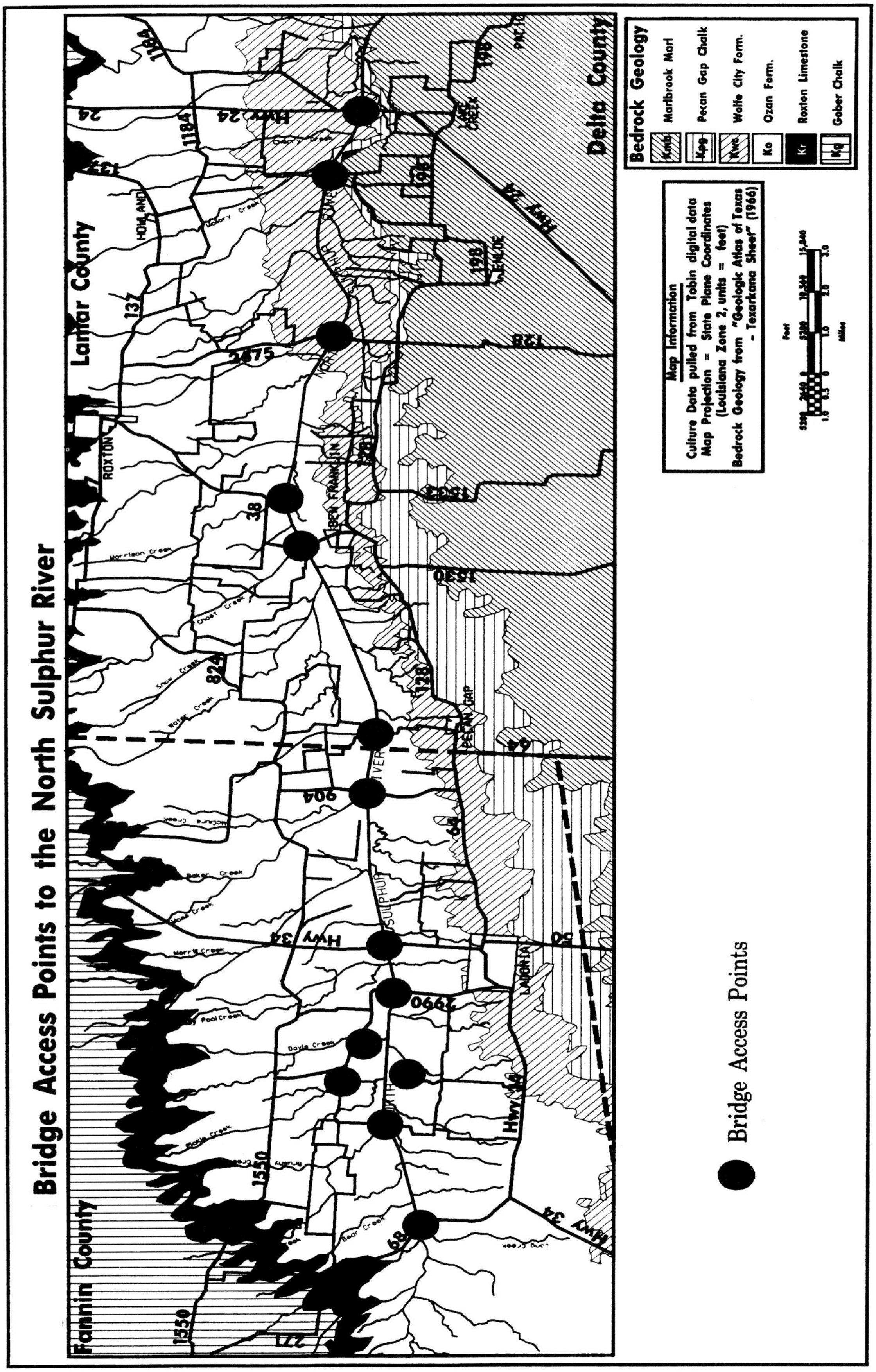

Fig. 1 Bridge access points along the length of the NSR study area.

The best time to collect is right after heavy rains have swept thru the area and the river level is already receding. Generally, a day or two after a heavy rainfall is sufficient time for it to be safe to go collecting in the river. Use caution and common sense. If you have ever collected the river before, you know how dangerous the flooding potential is. The banks are very tall and nearly vertical in most places. There is ample evidence of slumping and bank failure. As I mentioned at the beginning of the chapter, there are not many places to climb out of the river channel outside of the bridge crossings.

Do not go collecting on days when heavy rains are forecast in the area. Your field of vision is very limited in the river bottom. If it is raining real hard upstream from where you are, you will not know it till the water level starts to rise. And the water level rises fast in the NSR. All the gravel bars can be underwater in a matter of MINUTES! Always know where you are relative to an exit point out of the river. Failure to plan is planning to fail. Fossils are not worth drowning for!

The good news is that with our mild Texas winters you can collect the NSR all year round. In fact, winter collecting, especially in dry years, is the best time to collect. The temperature is not too hot and the water is not too cold. In the late summer the water level is at its lowest and the river channel looks more like a series stagnant puddles than a continuous body of water. There is usually not too much of a breeze down in the river channel proper. If it is 95 Fahrenheit and sunny up on top, then it will feel like 110 Fahrenheit down below with all the sunlight reflecting off the rocks and water. Dress accordingly and bring plenty of fluids to drink.

No matter how early you arrive to the NSR after heavy rainfalls, do not be disappointed by seeing a set of footprints in the river bottom in front of you. There must be ogres living under the bridges who like to collect fossils. The NSR is not a "secret" fossil locality. The NSR Fossil Symposium proved that. Collectors have been hitting it hard for thirty plus years. It is no longer the old days when people stacked up *Exogyra ponderosa* five feet tall because they were scattered everywhere. Don't you just hate hearing stories like that? Just remember, every time it rains something new and exciting has the potential of weathering out for the patient fossil collector to find.

Figs. 2 and 3 are distribution maps of collected Cretaceous macroinvertebrates along the NSR study area. The tabulations are based on the fossil material brought to the NSR Symposium. Naturally, fossil occurrences are clustered around the bridges crossing the NSR. Please note the abundance of molluscs, especially cephalopods, along the corridor extending from the Highway 34 bridge north of Ladonia upstream 1.2 miles from the FM 2990 bridge to the headwaters of Davis Creek. This coincides with the outcrop belt of the "red zone" within the Ozan Formation and immediately downstream from it. There is another productive area just downstream from the Highway 24 bridge in the eastern portion of the study area. This coincides with the beginning of the Pecan Gap Chalk outcrop belt on the NSR.

Fig. 4 is the distribution of Cretaceous shark remains found along the NSR. They are more evenly distributed along the length of the study area from the "red zone" outcrop belt to the downstream side of the Highway 24 bridge. A very productive locality is below and downstream from the railroad trestle over the river northeast from the town of Ben Franklin. This site is approximately 1.25 miles downstream from the Highway 38 bridge north of Ben Franklin. Shark's teeth can be found in the recent gravel bars and in the Pleistocene deposits along the riverbank on the south side.

Fig. 5 is the distribution of Cretaceous and Pleistocene vertebrates along the NSR study area. They have the most random and widespread distribution. There are multiple reasons for this. Most of these fossils are larger pieces easily seen by the naked eye while walking the gravel bars. Therefore they are collected more often than the smaller specimens. This indicates a collecting bias over smaller pieces that require kneeling, crawling or special methods to find.

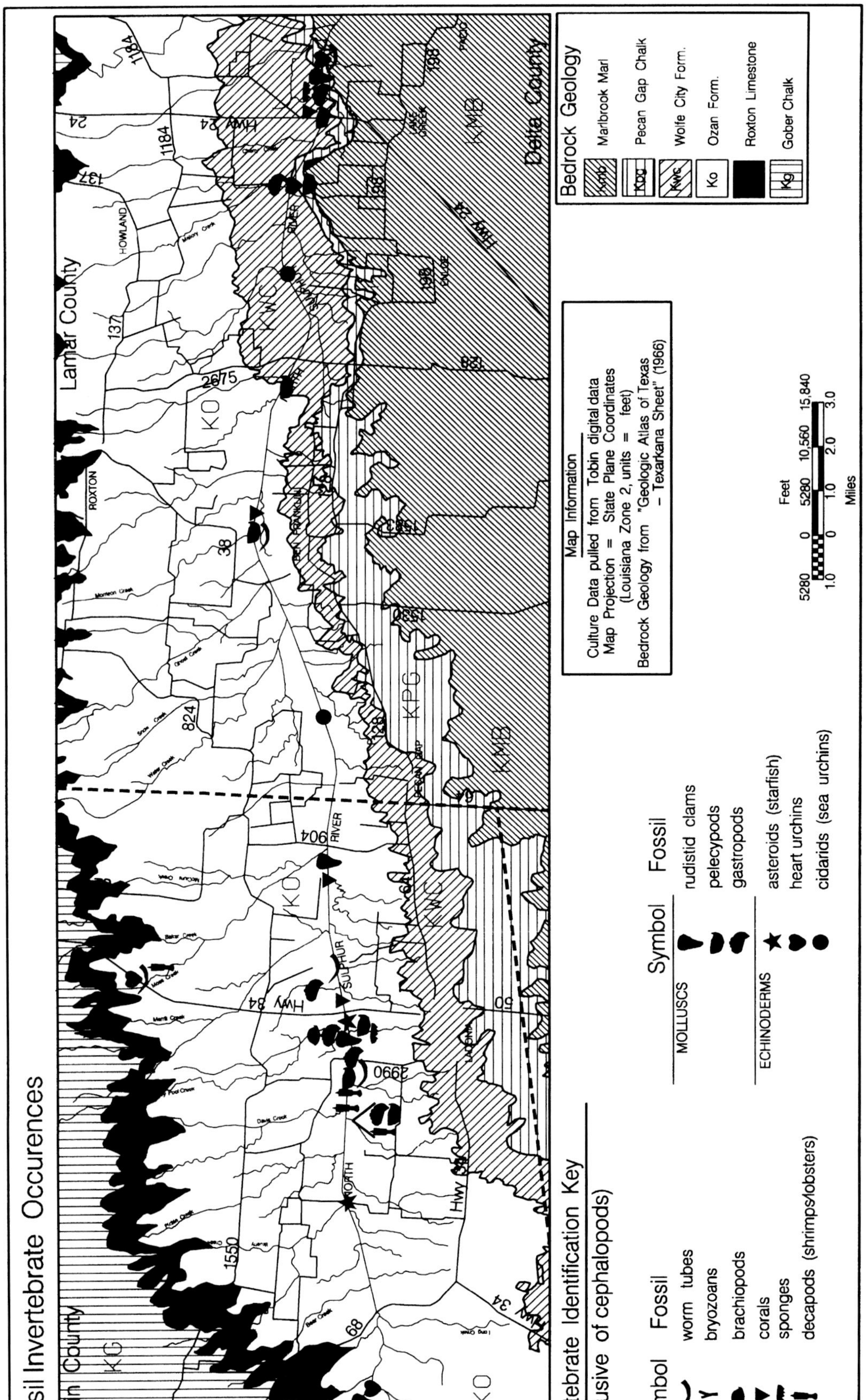

Fig. 2 Fossil invertebrate occurrences along the NSR study area.

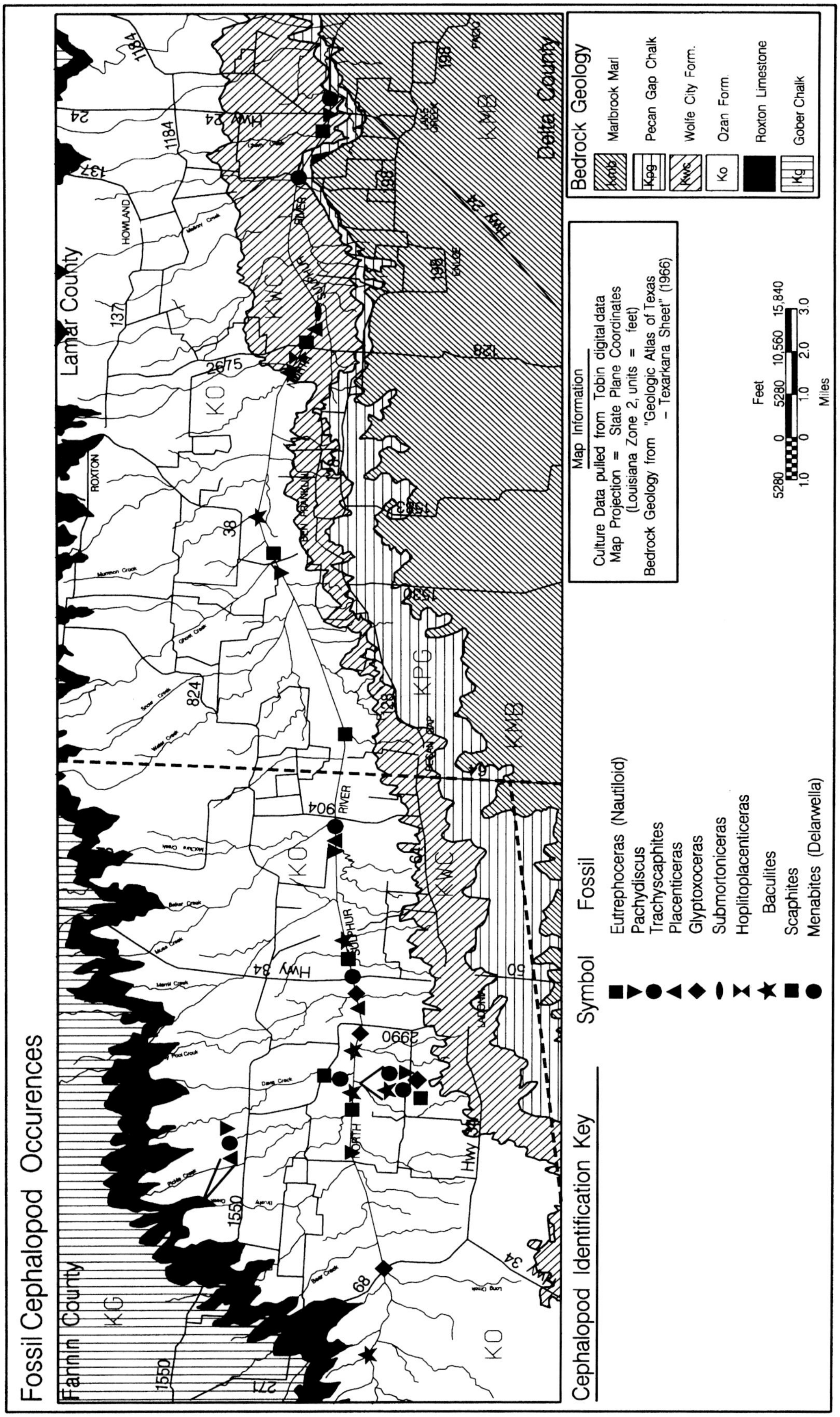

Fig.3 Fossil cephalopod occurrences along the NSR study area.

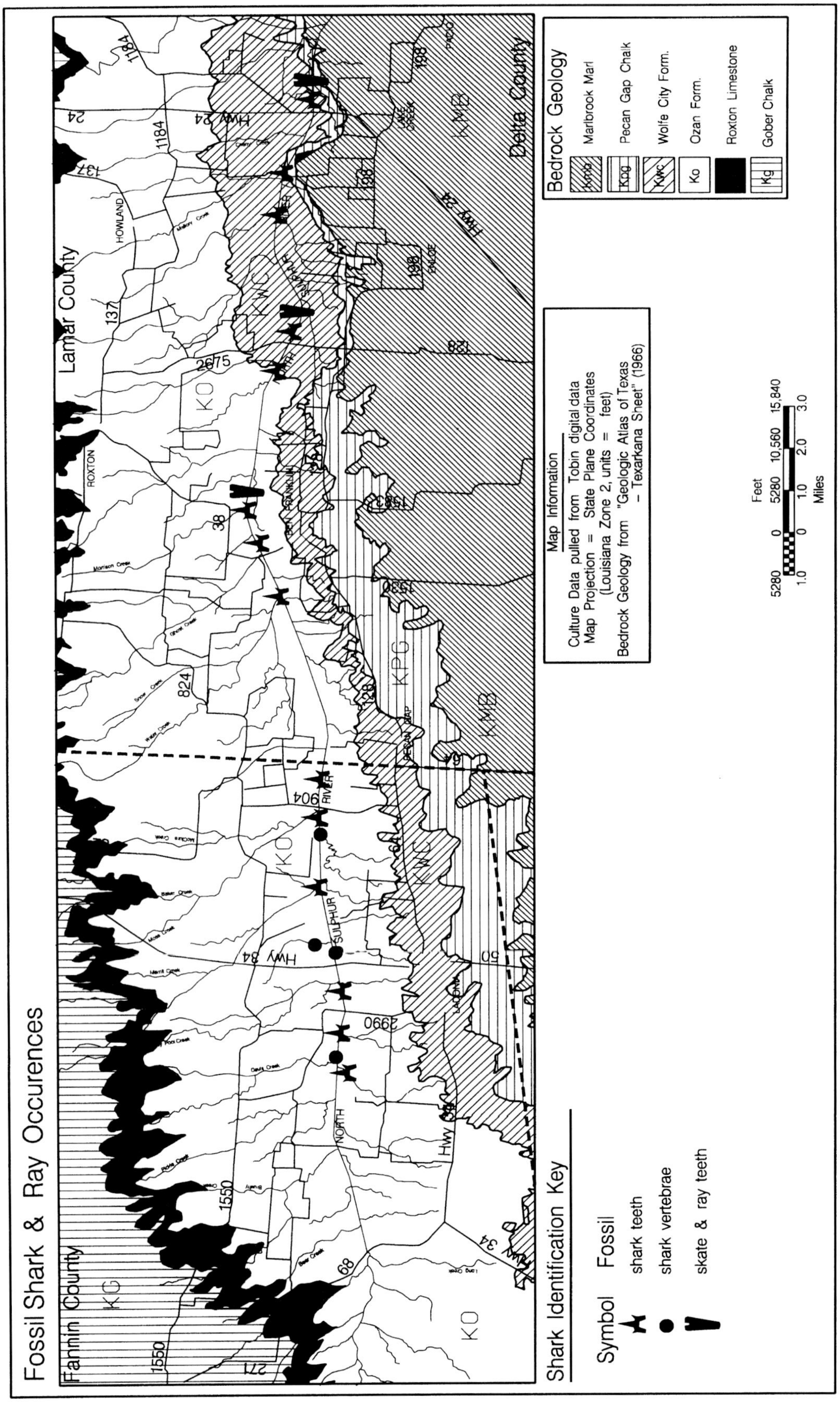

Fig. 4 Fossil shark and ray occurrences along the NSR study area.

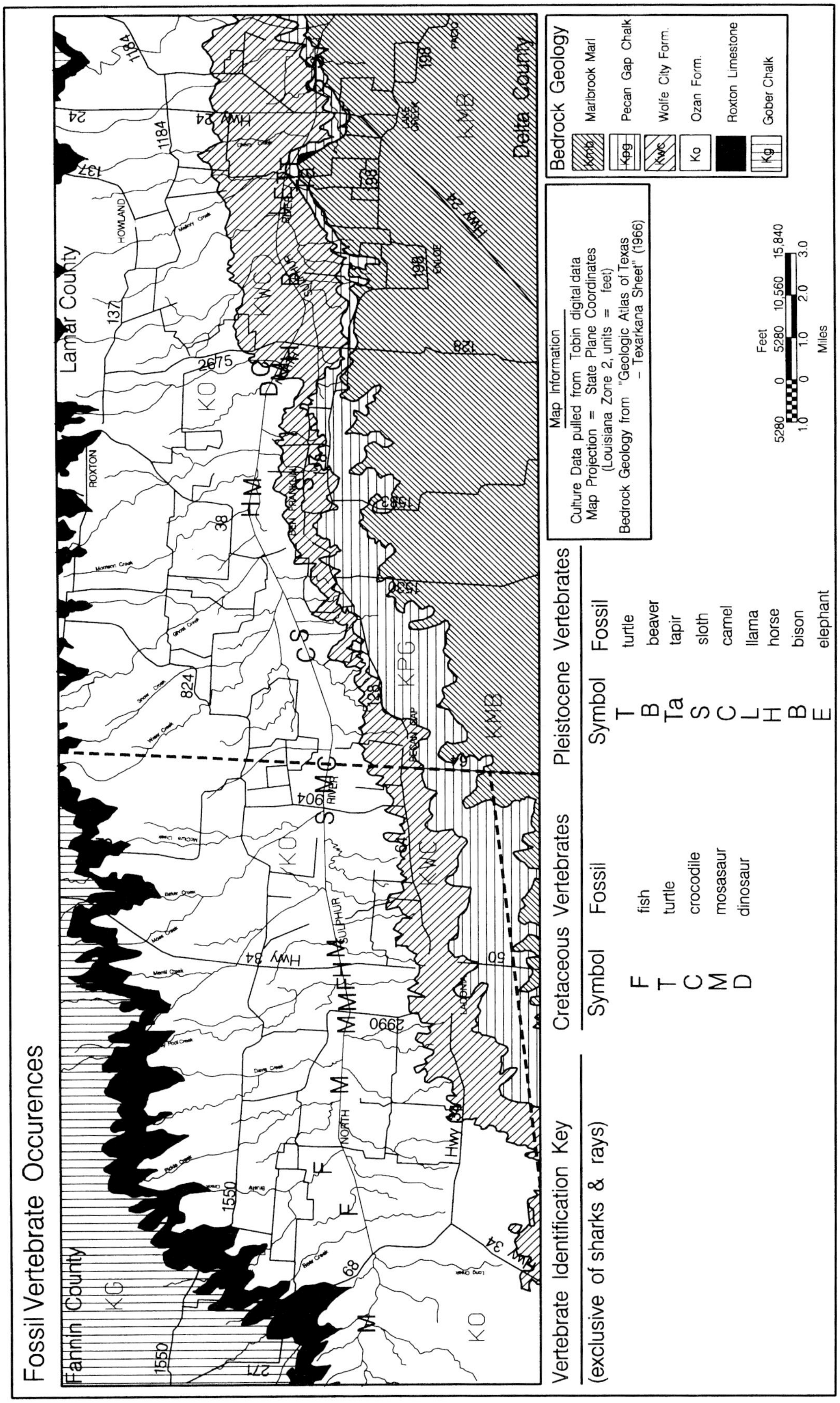

Fig. 5 Cretaceous and Pleistocene vertebrate occurrences along the NSR study area.

What information should you capture along with the specimen you are collecting? When we held the NSR Fossil Symposium, there were way too many people who came in with fantastic specimens who couldn't quite remember exactly where they had picked them up. And that is because they had collected it years previously and had collected thousands of other fossils in the meantime. We realize some sites are jealously guarded secrets that can only be revealed upon one's deathbed. We even have a few of them ourselves. The "if I tell you I must kill you" localities.

However, if you were to die tomorrow, do you have labels on the fossils in your collection? You should. And you don't want to burden your wife, husband or significant other with trying to memorize all the fossils in your den of antiquity. One collector's treasure is only a museum's study specimen (without a label). Mark McKinzie was recently involved in sorting through a very nice collection donated to a local college. Many of the fossil specimens were not labeled or obviously mislabeled. Some beautiful pieces could not be put on display because there was no other information to convey with them. A very sad situation.

Another example is attending a gem & mineral show. Occasionally you will see an exceptional specimen with no label information. It is hard to get interested if the dealer can't supply any details about it. How can you put a price on something if you don't even know what it is or where it came from?

There is a minimum amount of geographic, geologic, and paleontological information that you can record pertaining to that fossil. Fig. 6 is an example of the type of data you should capture while it is fresh in your memory. Get into the habit of recording this information when you get home from a collecting trip or after cleaning your fossils. It eventually becomes second nature and part of the routine. This is especially important if you find a new species of something that needs to be formally described. The professional describing it needs detailed locality information to cross check the locality against published geologic maps. He uses the stratigraphic information to check the fossil against all other known occurrences for that respective time interval and formation. This is the first step in establishing if it is a new species or not.

Most of the fossils collected from the NSR can be cleaned up with a minimal amount of preparation. Place the fossils in a tub of water and soak them overnight. Scrub them with an old toothbrush the next day. This procedure removes the clay and mud matrix. It works extremely well for the fossils collected from the gravel bars. A little more persistence is required for fossils hammered out of bedrock. Take them home and chip away the remaining matrix with small chisels or old screwdrivers. Finish them off by scraping away the last bits of matrix with a dental pick. Finally, attach a label to them or to the container you are storing them in. The following is an example label;

OYSTER (PELECYPOD):
Exogyra ponderosa
Age: Early Mid Campanian / Late Cretaceous
Form: Ozan Form./ Taylor Group
Loc: gravel bar 1.5 miles downstream from the
Hwy 34 Bridge, N.E. of Ladonia, Hunt Co., Tx.

Microfossil collecting requires a little more planning prior to hitting the field. First, what is your collecting objective? Is it to screen for microteeth or acquire microfossils from matrix? There are two good general areas along the NSR for collecting microteeth. One is the backside (downstream portion) of gravel bars within the river channel. The other is the Pleistocene gravels in the riverbanks such as at the Ben Franklin railroad bridge locality. In both cases, you need a bucket (5 gallon or smaller), a small shovel, a plastic tub and a screen.

The screen mesh from a household screen door is an appropriate size and can be purchased at a local hardware store. Cut it to the size of your plastic tub and attach it to a sturdy wooden frame. This can be in the 2 X 2 feet range.

Use the shovel to scoop up the small gravel from the gravel bar or a load full from the gravel in the riverbank. Place the wooden-framed screen on top of the plastic tub and dump the load onto it. Take this into calf-high water and place in the river. Gently shake the screen up and down and sideways while keeping it positioned over the plastic tub. Continue this procedure until no more material appears to be falling through the screen into the tub. Scan the contents of the screen for fossils and then dump it into the river. Take the tub and swish out the water a couple of times to remove any clay and mud floating in suspension. Place the remainder into the bucket to take home and look at under a stereo-microscope for the larger size microfossils. Once again, remember to record the locality information in the field or when you get home.

Bulk sampling of matrix for microfossils requires different methods and more equipment. However, I will discuss field sampling and sample preparation methods. The first thing to do on the outcrop is to choose an area with fresh exposures. You want samples that are unweathered and uncontaminated by recent vegetation. This usually involves removing a few inches or more of matrix before digging out a chunk of matrix you will keep. 500 grams (or approximately one pound) is an appropriate size. Place it in a clean container for storage and label the location and formation.

Allow the sample to air dry for a few days back at the laboratory (or house as the case may be).
The matrix now needs to be pulverized to liberate the microfossils. This can be done by using a mortar and pestle. Pound the sample with the pestle. Take care not to grind it as you may destroy the more delicate microfossils. The other method is to use a rock hammer and place the sample in a plastic baggie and "gently" pound the sample without breaking the baggie.

For the Pecan Gap Chalk, soak the disaggregated sample in a container of hydrogen peroxide for 24 hours. For the Ozan Formation, try a mixture of Dawn dishwasher detergent and water for 24 hours. Do not drain the liquid off and put the samples in a 3-speed Osterizer blender for 30 to 60 seconds, gradually increasing the speed from low to high. By the way, your margaritas will never taste the same again! Wash the blender mess through a 230-mesh stainless steel sieve to remove the clay size particles. Let the residue dry and view with an electron microscope. Hopefully, there will be all sorts of microfossils under the scope. If not, at least it made you feel like a mad junior scientist for the day.

Figure. 6 **Fossil Locality Information Datasheet:**

Part A – Geographic Information

Name:____
Date & Time:____
County/State/Country:____________________________________
General Location Information:________________________________

Part B – Geologic Information

1) Stratigraphy

1A) Bed:______________________________________
1B) Member:____________________________________
1C) Formation:__________________________________
1D) Group:____________________________________
1E) Age:______________________________________

Part C – Paleontological Information

1) Sampling Method:(quarried, sieved, surface collected)

____

2) Stratigraphic Position & Orientation within Bed:

2A) Distance from top of bed________________
2B) Orientation relative to bedding plane__________________
2C) "In situ" or float?__________________________

3) Taphonomic Information for **vertebrate** specimens:

3A) Bone occurrence: isolated (Y/N)_____________
associated (Y/N)____________
articulated (Y/N)____________

3B) For Bone Assemblages: concentrated or dispersed?:________________________
proximity of bones to one another?____________________
size range of bone material?___________________________

The following is a comprehensive bibliography pertaining to all aspects of the geology and paleontology of the North Sulphur River. It is current as of January 1999. The listing was painstakingly compiled over a two year period by Richard Zack. The Dallas Paleontological Society wanted to build an authoritative reference guide since the inception of the original North Sulphur River Project in 1996.

The reader is encouraged to use the following as a starting point for further research into the fossils of the North Sulphur River. This listing includes references both national and foreign, published and unpublished, common and obscure. It includes books, magazines, periodicals, special publications, and unpublished college level theses and dissertations. A majority of the material cited below will NOT be found at a public library. The reader should visit the geology department of the nearest college or university and see what resources are available. Most schools libraries have pay-for-copy reproduction machines onsite. Only an active student will be able to check out the reference materials.

The geology departments at Southern Methodist University (SMU) in Dallas and Texas Christian University (TCU) in Fort Worth have extensive geologic libraries. The DeGolyer Science Library at SMU is a very good source of information. Two good periodicals to start with are the *Journal of Paleontology* and the *Journal of Vertebrate Paleontology*. Books under the general heading of paleontology can be found in the bookshelves with the Dewey decimal numbers "560.xxx". An excellent entry-level book to the fossils of Texas is Charles Finsley's "*A Field Guide to Fossils of Texas*". Check for it in the regional section at bookstores such as Barnes and Noble or at the gift shop at the Dallas Museum of Natural History.

I encourage you to learn as much as possible about the geology of the NSR and the fossils you pick up in the field. It can be an addictive pursuit. Not to mention rewarding. And you will soon realize that the more you understand, the less you really know. Happy hunting!

North Sulphur River Geology/Paleontology Bibliography

(Sorted Alphabetically by Author)

Adkins, W.S. 1929. Some Upper Cretaceous ammonites from Texas. University of Texas Bulletin, 2901: 203-211.

Adkins, W.S. 1933. Geology of Texas Part 2; The Mesozoic systems in Texas. University of Texas Bulletin, 3232:239-518.

Adkins, W.S. 1936. Upper Cretaceous unconformities in Texas. University of Texas Bulletin, 3501:141-149.

Akers, R.E. and Akers, T.J. 1987. Texas Cretaceous Echinoids. Paleontology Section, Houston Gem and Mineral Society, 3:144 p.

Akers, R.E. and Akers, T.J. 1997. Texas Cretaceous Gastropods. Paleontology Section, Houston Gem and Mineral Society, 6:340 p.

Alexander, C.I. 1929. Ostracoda of the Cretaceous in North Texas. University of Texas Bulletin, 2907: 137 p.

Baii, K.M., Dauu, M., and Carson, L. 1990. Efficiency in collecting fossils. Proceedings of the Arkansas Academy of Sciences, 44:13-15.

Barrow, T.D., 1953, Mesozoic stratigraphy and structural history of the East Texas Basin. Stanford University unpublished phD dissertation, 221 p.

Beall, A.O., Jr. 1960. The Taylor Formation. Baylor Geological Society Field Conference (April 1960): 118-123.

Beall, A.O., Jr. 1964. Stratigraphy of the Taylor Formation (Upper Cretaceous), East – Central Texas. Baylor Geological Studies Bulletin, 6: 34 p.

Bottjner, J.D. 1979. Ichnology and depositional environments of Upper Cretaceous chalks, Southwest Arkansas (Annona Fm, Chalk Mbr. Saratoga Fm). Bulletin of the American Association of Petroleum Geologists, 63(4):422.

Bottjner, J.D. and Bryant,W.A. 1980. Paleoenvironmental analysis of disconformity and condensed bed at contact of Austin and Taylor Groups (Upper Cretaceous), East- Central and Northeastern Texas. Bulletin of the American Association of Petroleum Geologists, 64(5):679.

Bottjner, J.D. 1986. Campanian – Maastrichtian chalks of Southwestern Arkansas: petrology, paleoenvironments, and comparison with other North American and European chalks. Cretaceous Research, 7:161-196.

Bowman, T.E. 1971. *Palaega lamnas,* new species (Crustacean Isopoda) from the Upper Cretaceous of Texas. Journal of Paleontology, 45:540-541.

Burnhouse, M. 1958. Gulf Coast Tectonics. Bulletin of the American Association of Petroleum Geologists, 42(2):339-370.

Caron, M. 1983. A neotype for *Globotruncanita insignis* (Gandolfi). Journal Foraminifera Research, 13(2):144-145.

Cheatum, E.P. and Allen, D. 1963. An ecological comparison of the Ben Franklin and Clear Creek local molluscan faunas in Texas. Journal of the Graduate Research Center, 31:174-179.

Chimene, J.B. 1982. Occurrence of ostracoda species in the Upper Taylor Group (Upper Cretaceous) of central Texas. Maddock,R.F. (ed), Texas Ostracoda; Guidebook of Excursion and Related Papers for the 8th International Symposium on Ostracoda. University of Houston Department of Geosciences:206-210.

Chimene, J.B. and Maddock, R.F. 1984. Ostracode biostratigraphy and paleoecology of Upper Taylor Group (Campanian Stage – Upper Cretaceous) in central Texas. Bulletin of the American Association of Petroleum Geologists, 68(9):1210.

Clark, D.L. and Bird, K.J. 1966. Foraminifera and Paleoecology of the Upper Austin and Lower Taylor (Cretaceous) Strata in north Texas. Journal of Paleontology, 40:315-327.

Cobban, W.A. and Reedside, J.B., Jr. 1952. Correlation of the Cretaceous formations of the Western Interior of the United States. Bulletin of the Geological Society of America, 63:1011- 1044.

Cobban, W.A. 1969. The Late Cretaceous ammonite *Scaphites leei* (Reeside) and *Scaphites hippocrepis* (Dekay) in the Western Interior of the United States. United States Geological Survey Professional Paper, 619:1-57.

Cobban, W.A. 1973. The Late Cretaceous ammonite *Trachyscaphites pulcherrimus* (Roemer) in New Jersey and Texas. Journal of Research United States Geological Survey, 6(6):695-700.

Cobban, W.A. and Kennedy, W.J. 1992. Campanian ammonites from the Upper Cretaceous Gober Chalk of Lamar County,Texas. Journal of Paleontology, 66(3):440-454.

Cobban, W.A. 1992. Campanian *Trachyscaphites spiniger* ammonite fauna in Northeast Texas. Paleontology, 35(1):63-93.

Cobban, W.A. and Kennedy, W.P. 1993. Middle Campanian ammonites and inoceramids from the Wolfe City Sand in northeastern Texas. Journal of Paleontology, 67(1):71-82.

Cooke, C.W. 1953. American Upper Cretaceous echinoids. United States Geological Survey Professional Paper, 254-A:1-44.

Coon, L.A. 1956. Tertiary-Cretaceous growth of the East Texas Basin. Transactions of the Gulf Coast Association of Geological Society, 6:85-90.

Crosby, G.W. 1971. Gravity and mechanical study of the great bend in the Mexia-Talco Fault Zone, Texas. Journal of Geophysical Research, 76(11):2690-2705.

Cushman, J.A. 1946. Upper Cretaceous foraminifera of the Gulf Coast region of the United States and adjacent areas. United States Geological Survey Professional Paper, 206:241 p.

Dane, C.H. and Stephenson, L.W. 1928. Notes on the Taylor and Navarro formations. Bulletin of the American Association of Petroleum Geologists, 12(1):41-58.

Dane, C.H., 1929, Upper Cretaceous formations of southwestern Arkansas. Arkansas Geological Survey, 1:215 p.

Dawson, W.C. 1976. Petrography and sedimentation of the Upper Cretaceous Farmersville Member, Pecan Gap Formation, Taylor Group of Northeast Texas. University of Texas at Arlington unpublished Master's thesis, 158 p.

Dawson, W.C. and Reaser, D.F. 1976. Stratigraphy and depositional environment of the Upper Cretaceous Farmersville Member, Pecan Gap Formation, Taylor Group of Northeast Texas. Geological Society of America, Abstracts with programs, 8(1):16-17.

Dawson, W.C. and Reaser, D.F. 1994. Biogenic mounds and associated trace fossils: Wolfe City Formation (Upper Cretaceous), North-central Texas. Bulletin of the American Association of Petroleum Geologists, 78(9):1454.

Doering, J.A. 1963. Correlation of the Sulphur River Formation. Journal of the Graduate Research Center, 36(3):149-157.

Eaton, R.W. 1956. Resume of the subsurface geology of northeast Texas with emphasis on salt structures. Transactions of the Gulf Coast Geological Society, 6:79-84.

East Texas Geological Society. 1945. Field trip guidebook and road log (Greenville and Paris-Clarksville area) of northeast Texas. Pecan Gap, Wolfe City, and Anona Formations, Upper Cretaceous, 11 p.

Echols, J. 1972a. Biostratigraphy and reptile fauna of the Upper Austin and Taylor Groups (Upper Cretaceous) of Texas with special reference to Hunt, Fannin, Lamar, and Delta Counties in Texas. University of Oklahoma unpublished phD dissertation, 244 p.

Echols, J. 1972b. Biostratigraphy and reptile fauna of the Upper Austin and Taylor Groups (Upper Cretaceous) of northeast Texas. Society of Vertebrate Paleontology Annual Meeting 32nd Program. University of Nebraska (abstract):12.

Echols, J. 1974. Tectonic control of taphonomy of reptilian faunas from the Upper Austin and Taylor Groups (Upper Cretaceous) in northeast Texas. Annual Meeting of the American Association of Petroleum Geologists, Abstracts 6:29.

Echols, J. 1984. Ammonite zonation in condensed zone, Middle Ozan Formation (Taylor Group – Upper Cretaceous) in northeast Texas. Bulletin of the American Association of Petroleum Geologists, 68(4):473.

Edson, J.E. 1976. Palynolgy of the Upper Cretaceous strata in northeastern Texas. University of Tulane unpublished phD dissertation, 188 p.

Emerson, B.L.,Emerson, J.H., Akers, R.E., and Akers, T.J. 1994. Texas Cretaceous Ammonites and Nautiloids. Paleontology Section, Houston Gem and Mineral Society, 5:440 p.

Finsley, C. 1989. A field guide to Fossils of Texas. Texas Monthly Press, 190 p.

Frizzell, D.L. and Anderson, I.J. 1950. Diastems in the Pecan Gap Chalk of Travis County, Texas. Journal of Sedimentary Petrology, 20(1):55-59.

Frizzell, D.L. 1954. Handbook of Cretaceous foraminifera of Texas. University of Texas - Bureau of Economic Geology, Report of Investigations, 22:232 p.

Frye, J.C. and Leonard, A.P. 1963. Pleistocene geology of the Red River Basin in Texas. University of Texas – Bureau of Economic Geology, Report of Investigations, 49:48 p.

Fuhs, F.J. and Robinson, H.M. 1923. Structural and Stratigraphic data of northeast Texas petroleum area. Economic Geology, 18(8):709-731.

Furisch, F.T., Kennedy, W.J. and Palmer, T.J. 1981. Trace fossils at a regional discontinuity surface: the Austin/Taylor (Upper Cretaceous) contact in central Texas. Journal of Paleontology, 55(3):537-551.

Gartner, S., Jr. 1968, Coccoliths and related calcareous nannofossils from Upper Cretaceous deposits of Texas and Arkansas. University of Kansas – Paleontological Contributions, Article 1, 56 p.

Gordon, C.H. 1911. Geology and underground waters of northeastern Texas. United States Geological Survey, Water Supply Paper 276:1-12.

Granata, W.H.,Jr. 1963. Cretaceous stratigraphy and structural development of the Sabine Uplift area in Texas and Louisiana. Shreveport Geological Society, Report on selected north Louisiana and south Arkansas oil and gas fields, 5:50-95.

Hager, D.S. and Burnett, C.M. 1960. Mexia-Talco fault line in Hopkins and Delta Counties, Texas. Bulletin of the American Association of Petroleum Geologists, 44(3):316-356.

Hanger, R.A. and Echols, J. 1998. Nautiloids from the Cretaceous (Campanian) Ozan Formation, Taylor Group, Fannin County, Texas. The Texas Journal of Science, 50(3): 195-198.

Hardin, F.R. and Hardin, G.C.,Jr. 1961. Contemporaneous normal faults of the Gulf Coast and their relation to flexures. Bulletin of the American Association of Petroleum Geologists, 45(2): 238-248.

Hazel, J.E. and Paulson,O.L.,Jr. 1964. Some new ostracoda species from the Austin and Taylor (Coniacian and Campanian) rocks of the East Texas Embayment. Journal of Paleontology, 38(6):1047-1064.

Hill, R.T. 1894. Geology of parts of Texas, Indian Territory, and Arkansas adjacent to the Red River. Geologic Society of America Bulletin, 5:297-338.

Holman. J.A. 1963. Late Pleistocene amphibians and reptiles of the Clear Creek and Ben Franklin Local Fauna of Texas. Journal of the Graduate Research Center, 31:152.

Kauffman, E.G. 1977. Illustrated guide to biostratigraphically important Cretaceous macroFossils, Western Interior Basin, USA. Mountain Geologist, 14:225-274.

Kennedy, W.J. and Cobban, W.A. 1990. The Madagascar ammonite *Neogeothiericeras* Collignon, 1969 from the Upper Cretaceous (Campanian) of Texas. Palaont Z, 64:57-61.

Kennedy, W.J. and Cobban, W.A. 1991. Upper Cretaceous (upper Santonian) *Boehmoceras* fauna from the Gulf Coast region of the United States. Geological Magazine, 128(2): 167-189.

Kennedy, W.J., Cobban, W.A. and Scott,G.R. 1992. Ammonite correlation of the uppermost Campanian of western Europe, the U.S. Gulf Coast Atlantic Seaboard and Western Interior, and the numerical age of the base of the Maastrichtian. Geology, 129(4):497-500.

Kennedy, W.J. and Cobban, W.A. 1993. Campanian ammonites from the Annona Chalk near Yancy, Arkansas. Journal of Paleontology, 67(1):83-97.

Kennedy, W.J., Cobban, W.A. and Landman, N.H. 1997. Campanian ammonites from the Tombigbee Sand Member of the Eutaw Formation, the Mooreville Formation, and the basal part of the Demopolis Formation in Mississippi and Alabama. American Museum Novitates, 3201: 1-44.

Larson, N.L., Jorgensen, S.D., Farrar, R.A. and Larson, P.L. 1997. Ammonites and the other cephalopods of the Pierre Seaway. Geoscience Press,Inc., 148 p.

Leidy, J. 1865. Cretaceous reptiles of the United States. Smithsonian Contributions to Knowledge 192, 14(article 26):125 p.

Liddicoat, J.C., Hazel, J.E. and Brouwers, E.M. 1981. Magnetostratigraphy of Upper Cretaceous deposits in southwestern Arkansas and northeastern Texas. Bulletin of the American Association of Petroleum Geologists, 65(4):764-765.

Lillegraven, J.A. 1991. Stratigraphic placement of the Santonian-Campanian boundary (Upper Cretaceous) in the North American Gulf Coastal Plain and Western Interior, with implications to global geochronology. Cretaceous Research, 12:115-136.

Locke, K.A. 1995. Nannofossil biostratigraphy and magnetostratigraphy of the Taylor and Navarro Groups of north-central Texas. Texas A & M University unpublished phD dissertation, 289 p.

Lokke, D.H. and Brent, M.D. 1966b. Pleistocene – Recent chalk gravel deposits of Fannin County, Texas. Compass, 43(2):134-142.

McNulty, C.L., Jr. 1955. Foraminifera of the Austin Group in northeast Texas. University of Oklahoma unpublished phD dissertation, ? p.

McNulty, C.L., Jr. 1962a. Some observations on Austin equivalent in northeastern Texas. Texas Journal of Science, 14(4):418.

McNulty, C.L., Jr. 1976. *Globotruneata fornicata* zone of Upper Austin Group (Cretaceous), northeast Texas. Bulletin of the American Association of Petroleum Geologists, 60:2058-2062.

McNulty, C.L., Jr. 1977. Emandation of the Pecan Gap Chalk (Campanian) of northeastern Texas. Transactions of the Gulf Coast Association of Geological Society, 27:430.

McNulty, C.L., Jr., Brezina, J.L., Dawson, W.C. and Malv, F.W. 1981. Emandation of the Pecan Gap Chalk (Campanian) of northeastern Texas. Transactions of the Gulf Coast Association of Geological Society, 31:353-358.

Morgan, B.E. 1962. Foraminifera of the Pecan Gap Formation (Cretaceous) in northeastern Texas. University of Oklahoma unpublished Master's thesis,? p.

Murray, G.F. 1961. Geology of the Atlantic and Gulf Coastal Province of North America. Harper and Brothers Publishing, New York, 692 p.

Nicholls, E.L. and Russell, A.P. 1990. Paleobiogeography of the Cretaceous Western Interior Seaway of North America: the vertebrate evidence. Paleogeography, Paleoclimatology, PaleoEcology, 79:149-179.

Nufer, J.A. 1977. Foraminifera of the Pecan Gap Chalk, their relationship to biostratigraphic contacts and depositional environment. Baylor University unpublished Master's thesis, 78 p.

Pate, K.A. 1996. The significance of the fossil communities of the Wolfe City Sand of the Taylor Group (Upper Cretaceous, Gulfian Series), central Texas. Bulletin of the Geological Society of America, 28(1):58.

Paulson, O.W.,Jr. 1960. Ostracoda and stratigraphy of Austin and Taylor equivalents of northeast Texas. Louisiana State University and A & M College unpublished phD dissertation, ? p.

Pessagno, E.A.,Jr. 1967. Upper Cretaceous planktonic foraminifera from the Western Gulf Coastal Plain. Paleontolographica Americana, 5(37):245-445.

Pessagno, E.A.,Jr. 1969. Upper Cretaceous stratigraphy of the western Gulf Coast area of Mexico, Texas, and Arkansas. Geological Society of America Memoir, 111:139 p.

Presley, O.D. 1965. Ostracoda of the Pecan Gap (Cretaceous) Formation of northeastern Texas. University of Oklahoma unpublished Master's thesis, ? p.

Rainey, M. 1974. The Quaternary stratigraphy of the North Sulphur River. Southern Methodist University unpublished Master's thesis, ? p.

Richardson, J.D. 1972a. Stratigraphy and depositional environment of the Wolfe City Formation (Upper Cretaceous), north-east Texas. University of Texas at Arlington unpublished Master's thesis, ? p.

Richardson, J.D. 1972b. Paleoecology of ichnofossils from the Wolfe City Formation (Upper Cretaceous), northeast Texas. Geological Society of America Abstracts with Programs, 4(4): 290-291.

Richardson, J.D. 1975. Trace fossils from the Wolfe City Formation (Upper Cretaceous), northeast Texas. Texas Journal of Science, 26(3/4):339-352.

Ronalder, N.L. 1982. Geologic investigation of the Austin-Taylor contact, northeast Texas. University of Texas at Arlington unpublished Master's thesis, 196 p.

Rouse, J.T. 1944. Correlation of the Pecan Gap, Wolfe City, and Annona Formations in east Texas. Bulletin of the American Association of Petroleum Geologists, 28(4):522-530.

Russell, D.A. 1970. The vertebrate fauna of the Selma Formation of Alabama, part 7: the mosasaurs. Chicago Museum of Natural History Fieldiena Geological Memoir, 3(7):365-380.

Russell, D.A. 1988. A checklist of North American marine Cretaceous vertebrates including fresh-water fishes. Tyrrell Museum Paleontological Occasional Papers, 4:1-58.

Russell, D.A. 1993. Evolution of the Western Interior Basin - Vertebrates in the Cretaceous Western Interior Sea. Geological Association of Canada Special Paper, 39:665-680.

Schlichting, H.E.,Jr. 1963. Charophytes of Pleistocene Age from Delta and Denton Counties, Texas. Journal of the Graduate Research Center, 31:180-181.

Seewald, K.O. 1967. Stratigraphy of the Upper Cretaceous Austin Group, central Texas. Transactions of the Gulf Coast Association of Geological Society, 17:151.

Sellers, E.H. and Baker, C.L. 1934. The Geology of Texas, volume 2, structural and economic geology. University of Texas Publication, 3401:884 p.

Sheu, N.W. 1982. Calcareous nannofossils of the Brownstone Marl and Gober Chalk (Upper Cretaceous) of the Upper Austin Group, Fannin County, northeast Texas. University of Texas at Arlington unpublished Master's thesis, ? p.

Sisk, C.T.,Jr. 1958. The Taylor Marl. Baylor Geological Society Guide to the Mid-Cretaceous geology of central Texas, 79-83.

Slaughter, B.H. and Hoover, B.R. 1963. Sulphur River Formation and the Pleistocene mammals of the Ben Franklin local fauna. Journal of the Graduate Research Center, 36:132-148.

Slaughter, B.H. and Ritchie, R. 1963. Pleistocene mammals of the Clear Creek local fauna, Denton County, Texas. Journal of the Graduate Research Center, 31:117-131.

Slaughter, B.H. 1964. Geological survey and appraisal of the paleontological resources of the Cooper Reservoir Basin, Delta and Hopkins Counties, Texas. Fondren Sciences Service, 6:11 p.

Stehli, F.G., Creath, W.B., Upshaw, C.F. and Forgotson, J.M. 1972. Depositional history of Gulfian Cretaceous of the East Texas Embayment. Bulletin of the American Association of Petroleum Geologists, 56(1):38-67.

Stephenson,L.W. 1918. A contribution to the geology of northeastern Texas and southeastern Oklahoma. United States Geological Survey Professional Paper, 120-H:129-163.

Stephenson, L.W. 1927. Notes on the stratigraphy of the Upper Cretaceous formations of Texas and Arkansas. Bulletinof the American Association of Petroleum Geologists, 2(1):1-17.

Stephenson, L.W. 1929. Unconformities in the Upper Cretaceous series of of Texas. Bulletin of the American Association of Petroleum Geologists, 13:1323-1334.

Stephenson, L.W. 1936. Stratigraphic relations of the Austin, Taylor, and equivalent formations in Texas. United States Geological Survey Professional Paper, 186-G:133-146.

Storrs, G.W. 1981. A review of the occurrence of the plesiosauria (Reptilia; Sauropterygia) in Texas. University of Texas at Arlington unpublished Master's thesis, 226 p.

Thomas, N.L. and Rice, E.M. 1931. Cretaceous chalks, Texas and Arkansas. Bulletin of the American Association of Petroleum Geologists, 15:965-966.

Thompson, L.B., Percival, S.F.,Jr. and Patricelli, J.A. 1978. Stratigraphic relations of the Annona Chalk and Gober Chalk (Upper Campanian); type localities in northeast Texas and southwest Arkansas. Transactions of the Gulf Coast Association of Geological Society, 28:665-679.

Thompson, L.B., Heine, C.J., Percival, S.F.,Jr. and Selznick, M.R. 1991. Stratigraphy and micropaleontology of the Campanian shelf in northeast Texas. Micropaleontology Special Publication, 5:148 p.

Thurmond, J.T. 1969. Notes on mosasaurs from Texas. Texas Journal of Sciences, 21(1):69-80.

Tucker, D.R. and Hencey, D.P. 1987. Mini-shelves in Taylor shale: stratigraphy of Late Cretaceous and Early Tertiary units, Bastrop County, Texas. Bulletin of the American Association of Petroleum Geologists, 71(9):1123.

University of Texas – Bureau of Economic Geology. 1966. Geologic Atlas of Texas – Texarkana sheet. University of Texas – Bureau of Economic Geology, scale 1:250,000.

Uyeno, T. 1963. Late Pleistocene fishes of the Clear Creek and Ben Franklin local faunas of Texas. Journal of the Graduate Research Center, 31:168-173.

Walthall, B.H. and Wolper, J.L. 1967. Peripheral Gulf riftings in northeast Texas. Bulletin of the American Association of Petroleum Geologists, 51(1):102-110.

Waters, J.A. , McFarland, P.W. and Lea, J.W. 1955. Geologic framework of Gulf Coastal Plain of Texas. Bulletin of the American Association of Petroleum Geologists, 39(9):1821-1850.

Welton, B.J. and Farish, R.F. 1993. The collector's guide to fossil sharks and rays from the Cretaceous of Texas. BeforeTIME Publishing, 204 p.

Williston, S.W. 1900. Cretaceous fishes. University of Kansas Geological Survey, 6:235-286.

Young, K.P. 1963. Upper Cretaceous ammonites from the Gulf Coast of the United States. University of Texas Publication, 6304:373 p.

Young, K.P. 1963. A revision of Taylor nomenclature, Upper Cretaceous. University of Texas – Bureau of Economic Geology Geologic Circular, 65(3):10 p.

Glossary of Terms

aff. - Affinity. Used to indicate a species closely related to, but not identical to, a previously described species.

anterior- Front

anterolaterals- A rowgroup of teeth in which the similarities between anterior and lateral teeth make distinguishing between the two nearly impossible.

aperatural- View from the open end of a chambered body such as a cephalopod.

apical- View seen when looking at the tip of the crown of the tooth from above.

attachment surface- The portion of the root that seats in the dental membrane against the jaw surface.

basal- View seen when looking at the bottom of an organism from below.

basal ledge- The ledge formed by the expansion of the crown foot above the root.

benthic- The bottom of the ocean and the fauna dwelling there.

bilobate- Root having 2 lobes, distal and mesial.

cf.- Comparable to a described species.

coprolite- Fossilized excrement. The non-digestible portion of food passed through the digestive tract of an animal.

craton- Stable structural portion of the earth's continental crust.

crown- The enameloid-covered portion of a tooth.

cusp- The major crown projection of a tooth.

cusplet- A minor crown projection flanking the crown of a tooth.

dentary- The lower tooth-bearing jaw bone. It is the main bone forming the lower jaw in reptiles and the only one forming the lower jaw in mammals.

diagentic- The processes affecting rock layers and mineral composition at or near the surface.

digit- Single finger or toe.

distal- The opposite of mesial. Toward the hinge of the jaw.

dorsal- Toward the top.

dorsoventrally- From top to bottom.

epifauna- Organisms which live attached to other, generally larger organisms. The attachment may or not be symbiotic or beneficial to either organism.

extant- Still in existence.

extinct- No longer in existence. Commonly used in reference to an entire species and not an individual organism.

facies- The sum total of all features composing a rock unit. These include sedimentary rock type, bedding features, fossil content, etc.

fluvial- Depositional processes and products caused by rivers and streams.

foramen- Hole.

genus- A taxonomic term for a group of related species that have morphologic characteristics uniting them.

geomorphology- The description and interpretation of surface landforms.

ichnofossil- All the various tracks and trails left by fossil organisms in the rock record. Also known as trace fossils.

ichnogenus- An organism described on footprint data only.

index fossil- A fossil species which characterizes a specific time horizon by its abundance. May or may not be restricted to that horizon.

infaunal- Organisms which live beneath the surface. Generally refers to marine organisms living beneath the sediment/water surface.

in situ- (Latin; 'in place'). Fossils found in their original position within a rock formation.

interglacial- A period of glacial retreat is referred to as an interglacial (also known as 'interstadial').

labial- The face of a tooth facing the outside of the mouth.

labiolingual- The orientation of a tooth relative to the outside and the inside of the mouth.

lingual- The face of the tooth facing the tongue.

marine regression- Withdrawl of the seas from the continental shelves resulting in a net gain in exposed landmass.

marine transgression- Inward movements of the seas over the low lying shelves of the continental plates.

mesial- Opposite of distal. Toward the front of the jaw.

mesodistal- Referencing front to back of the jaw.

niche- The role played by each organism within its community. Generally used in the context of its feeding habits and lifestyle.

NSR- North Sulphur River.

nutrient groove- A groove separating the root lobes on the basal, or lingual root face.

occlusal- Same as apical.

oral- Teeth found in the mouth.

outcrop- Total area over which a particular rock unit occurs on the surface. Often used as a synonym for rock exposure.

pelagic- Lifestyle adopted by marine organisms living in the open sea but not on the sea floor, includes planktonic and nektonic forms.

Phanerozoic- (Latin; 'obvious life'). The period of time during which sediments accumulated that contain obvious remains of plants and animals. Typically refers to the entire time interval from the base of the Cambrian to present day.

phylogeny- The evolution of a related group of organisms through geologic time.

planktonic- Organisms which float passively with the currents in a body of water.

remanie- Organic skeletal material that has accumulated over a considerable period of time and has been subjected to transportation and abrasion prior to burial. Many bone beds and condensed sections contain fossils produced by these processes.

rostral- Spine-like teeth found along the edges of the flattened rostrum of sawfishes.

rostrum- Elongated snout of sawfishes and other billfish.

sigmoid- Having a shape resembling the letter "S".

sp.- Abbreviation for species.

species- From a biological standpoint, members of an animal or plant population that can interbreed. Fossil species are characterized by similar sets of morphological traits.

styolite- An irregular suture-like boundary that can develop in carbonates or evaporites and is caused by pressure induced solution and subsequent redeposition. Generally independent of bedding plane surfaces.

teleost- The largest and most advanced group of the ray-finned bony fishes, includes most living fishes.

tetrapod- A four-legged vertebrate animal.

var.- Abbreviation for variety or variation. Taxonomic term applied to very similar species or subspecies.

ventral- Underside of an object or organism.

WIS- Western Interior Seaway.

Index

T

U

W

X

Y

Fig. 1 View looking upstream from the FM 2990 bridge north of Ladonia.

Fig. 2 View looking downstream from the FM 2990 bridge north of Ladonia.

Fig. 3 Gober Chalk/basal Ozan Formation contact at the Highway 34 bridge over Moss Creek north of Bugtussle.

Fig. 4 Basal Ozan Formation (red) overlying the Gober Chalk (white) at Moss Creek.

Figs. 5, 6 & 7 Three views of the "red zone" within the Ozan Formation near the entrance of the unnamed creek across from Davis Creek 1.2 miles upstream from the FM 2990 bridge. Note the nodular appearance.

Fig. 8 The Pecan Gap chalk downstream from the Hwy 24 bridge at the Slabtown locality along a fault plane.

Fig. 9 Bank exposure of the Pecan Gap Chalk along Hickory Creek north of the river.

Fig. 1 *Zoophycus* feeding traces in a "red zone" nodule from the Ozan Formation. Mark McKinzie.

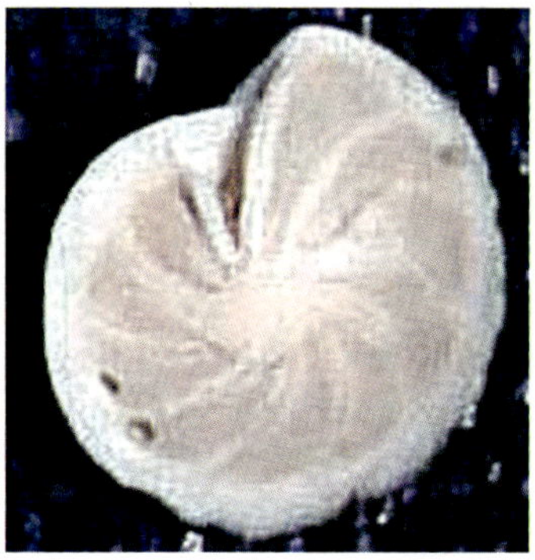

Fig. 2 Foraminifera *Lenticularia sp.* from the basal Ozan Formation. Ed Swiatovy. (x20)

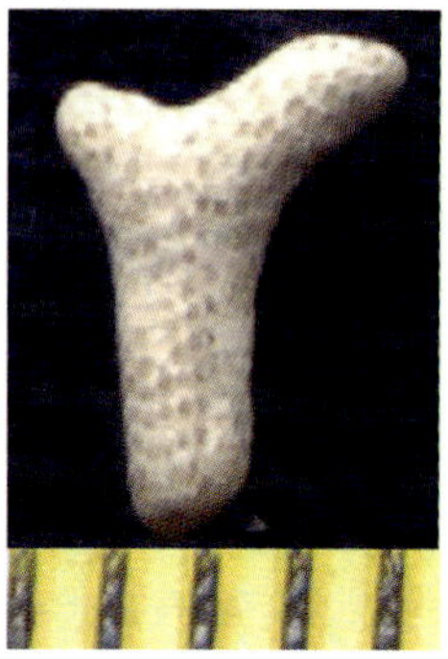

Fig. 3 Unidentified stick-like bryozoan colony. Scale bar in mm.

Fig. 4 Calcareous worm tubes of *Hamulus squamosus* from the Pecan Gap Chalk. Rick Kaiser.

Fig. 5 Individual plates of the barnacle *Acroscapellum sp.* Rick Kaiser.

Fig. 6 Terbratulid brachiopod *Choristothyris plicata*. Brian Bowles.

Fig. 7 Solitary "horn" coral *Trochocyathus sp.* from the Ozan Formation. Mac Glaess.

Figs. 8, 9 Distal and close up views of a large phaceloid coral colony from a nodule in the Pecan Gap Chalk. Possibly *Cladophyllia sp.*? Mac Glaess.

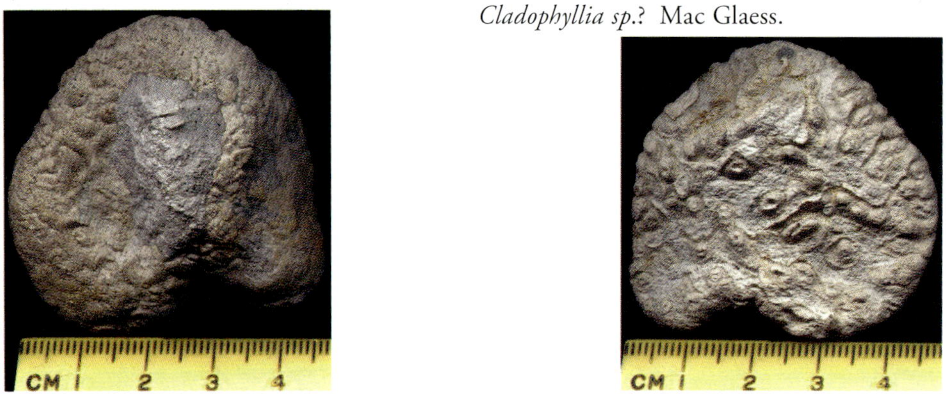

Figs. 10, 11 Dorsal and ventral views of the boring calcareous sponge *Cliona cretacia* from the Pecan Gap Chalk. The sponge itself is rarely collected. The only evidence usually preserved are the tubular borings found in pelecypod valves, especially *Exogyra ponderosa*. Mark McKinzie.

Fig. 1 Ventral (top) and dorsal (bottom) views of the carapace of *Linuparus sp.* from the "red zone" of the Ozan Formation. Dennis Magar.

Fig. 2 "Finger" claw segment of the lobster *Endoplyctia kimzeyi*? from the "red zone" of the Ozan Formation. Valerie Cheshier.

Fig. 3 Abdomen and tail fan of a lobster, probably *Hoploparia sp.*? from the Ozan Formation. Linda Farish.

Fig. 4 Extremely large carapace of unidentified crab from the Ozan Formation. Valerie Cheshier.

Fig. 5 Close-up detail of the carapace in Fig. 4.

Fig. 6 Two oral side up arm fragments of an unidentified astropectinid starfish. Tom Newman.

Fig. 7 Adoral (dorsal) view of a nearly complete and enrolled starfish from the Ozan Formation. Preservation of the specimen makes identification difficult.

Fig. 8 Oral (ventral) view of the same specimen in Fig. 7. Probably a new and undescribed species.

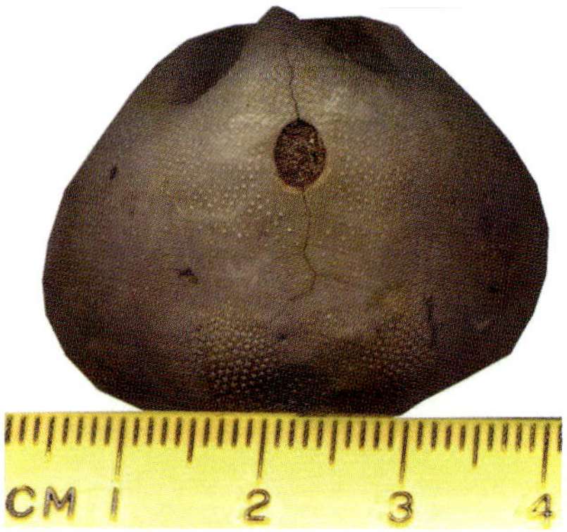

Figs. 1, 2, 3 Apical (top), oral (bottom) and anal (posterior) views of a well preserved test of the heart urchin *Hemiaster beecheri* from the basal unit of the Ozan Formation at Moss Creek. Mark McKinzie.

Fig. 4 Slab with multiple *Hemiaster beecheri* echinoids

Fig. 5 Badly weathered, unidentified regular echinoid, possibly *Rachiosoma sp.* from the Ozan Formation. Phil Kirchoff and Yvette Nethery.

Figs. 6, 7 Side and apical (top) views of a very well preserved heart urchin *Echinocorys texanus* from the Pecan Gap Chalk. Mac Glaess.

Figs. 8, 9 Oral (bottom) and apical (top) views of the regular echinoid *Pedinopsis pondi* from the Ozan Formation. Barbara Elliot.

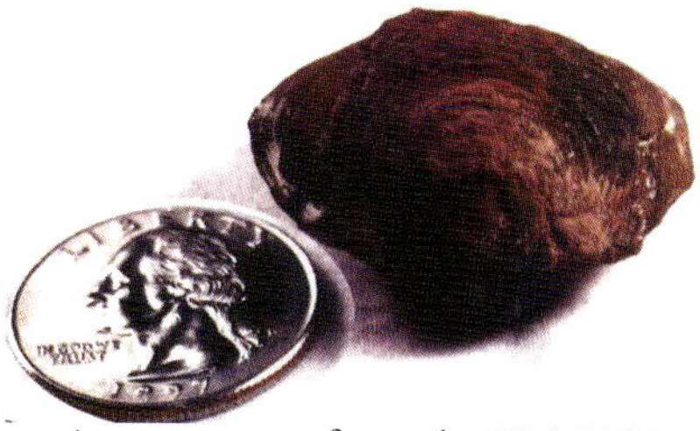

Figs. 1, 2, 3 Pelecypod *Anomia argentaria* from the "red zone" of the Ozan Formation upstream from the FM 2990 bridge. Fig.1 is two phosphatic molds. Figures 2 & 3 are two views of the another individual. All specimens Chris Vencevich.

Figs. 4, 5, 6 Three views (top, posterior and side) of a large well preserved specimen of *Pholodomya lincemani* from the Pecan Gap Chalk downstream from the Hwy 24 bridge. Mac Glaess.

Figs. 7, 8 Ventral(domed) and dorsal(flattened) valves of the large oyster *Exogyra ponderosa*. This fossil can be collected from the entire length of the NSR study area and is especially common in the Ozan and Pecan Gap Formations. Chris Vencevich.

Figs. 9, 10, 11 Three different specimens of an unusual rudist clam. Fig. 9 is a colonial form of *Sauvagesia belti* from the Pecan Gap Chalk downstream from the Hwy 24 bridge. Mac Glaess. Four separate individuals have grown together and resemble a coral colony. Fig. 10 is a highly weathered specimen of a very large *Durania (Radiolites) austinensis* from the Ozan Formation. Gerry Hightower estate collection. Specimens this large are commonly mistaken for petrified wood. Fig. 11 is a close up view (x2) of a portion of weathered *Sauvagesia belti* from the Ozan Formation. Mark McKinzie. Once again, notice how it resembles a colonial coral.

Fig. 1 Internal phosphatic molds of various gastropods. This is the normal type of preservation of gastropods from the NSR study area. The two gastropods in the upper row are *Gyrodes major*, a form genus. The two long slender specimens are probably *Turritella trilira*.

Fig. 2 Extremely well preserved specimen of *Turritella trilira* from the "red zone" of the Ozan Formation. Brian Bowles.

Fig. 3 Two specimens of *Anchura substriata* from the "red zone" of the Ozan Formation. Note the coarse ribbing. Mark McKinzie.

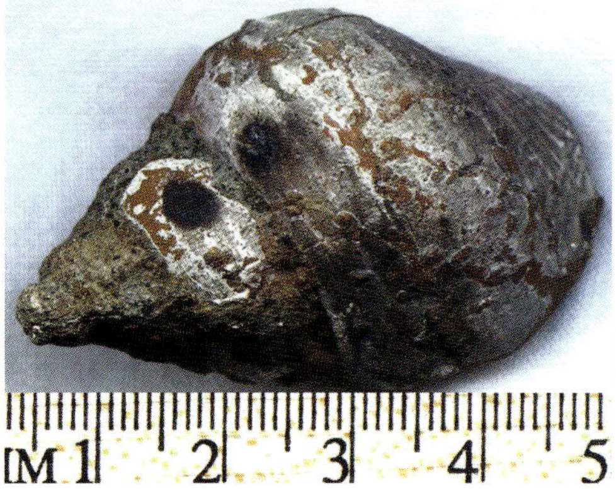

Figs. 4, 5 Two views of *Polinices (Euspira) rectilabrum* from the "red zone" of the Ozan Formation at the mouth of Davis Creek. Ornamentation consists of fine striations parallel to the shell coiling. Mark McKinzie.

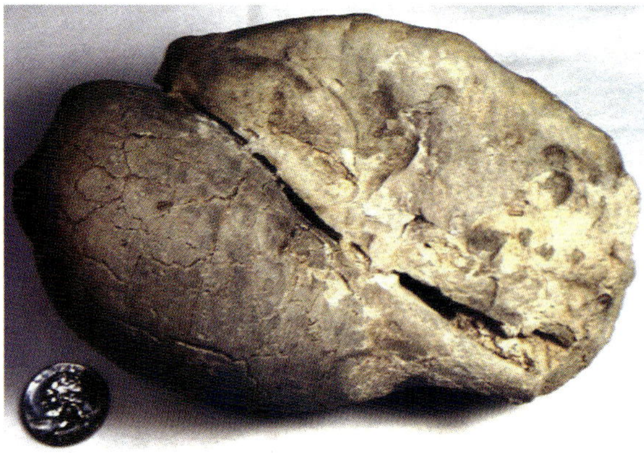

Figs. 6, 7 Two views of an extremely large internal mold of *Volutomorpha* aff. *conradi*. None of the original shell is present and the shell ornamentation is unknown. This is the largest genus of gastropod collected along the NSR study area. Lonnie French.

Figs. 1, 2, 3 Top, side, and apertural views of the coiled nautiloid *Eutrephoceras dekayi* from the Ozan Formation. This cephalopod can be distinguished from all the others by the simple sinuous sutures separating the body chambers. Tom Newman.

Fig. 4 Partial living chamber of a very large *Baculites sp. aquilaensis* from the "red zone" of the Ozan Formation. Chris Vencevich.

Fig. 5 Phosphatic internal mold of a Baculites sp.aquilaensis from the Ozan Formation.

Fig. 6 Whorl segments of the open large coiled heteromorph *Glyptoxoceras sp.* from the "red zone" of the Ozan Formation. Mark McKinzie.

Fig. 7 Juvenile specimen in matrix of *Hoplitoplacenticeras minor* from the Wolfe City Sandstone. Bill Jolley.

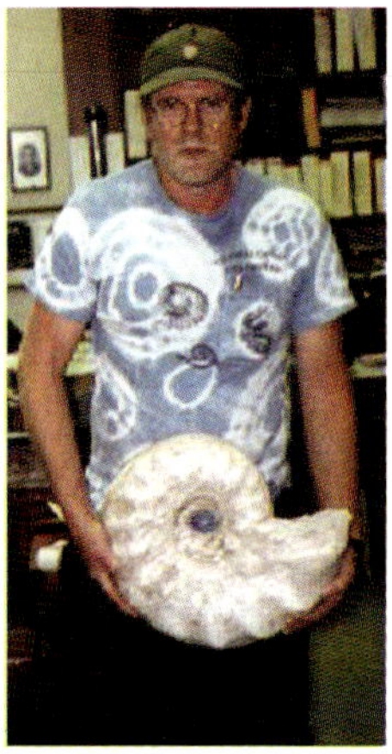

Figs. 8, 9 Two views of a very large *Hoplitoplacenticeras plasticum*? from the Wolfe City Sandstone. A standing Bill Jolley for scale.

Fig. 10 Large discoidal internal mold of *Placenticeras planum* from the upper Gober Chalk Formation. Bob Wilk.

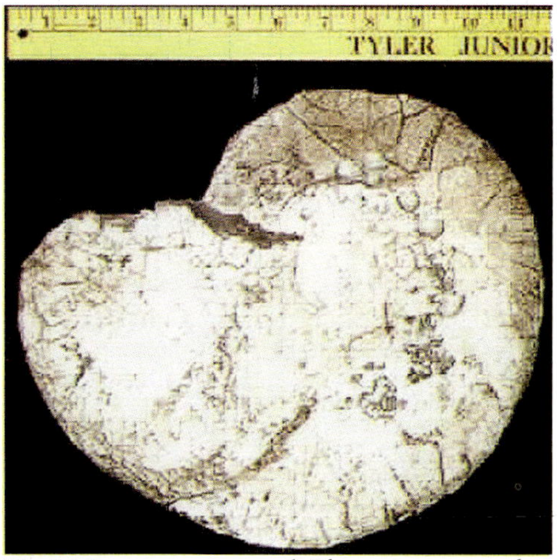

Fig. 11 Large compressed *Placenticeras syrtale* from the Wolfe City Sandstone. Bill Jolley.

Figs. 1, 2 Two well preserved specimens of *Texanites sp* from the Wolfe City Sandstone. *Texanites* can be distinguished from other similar sized ammonites by the presence of strong ribbing with 5 tubercles. Bill Jolley. This genus is more common in the underlying Austin Group outside of the NSR study area.

Figs. 3, 4, 5 Fig. 3 is a specimen of *Menabites (Delawarella) vanuxemi* from the Wolfe City Sandstone. Figs. 4 & 5 are two different specimens of *Menabites (Delawarella) delawarensis* from the Gober Chalk. Figs. 3 & 4 by Bill Jolley. Fig. 5 by Bob Wilk.

Fig. 6 Large and well preserved *Menabites (Delawarella) danei* from the "red zone" of the Ozan Formation upstream from the FM 2990 bridge. Notice the coarser ribbing *(Delawarella) delawarensis* from the "red zone" of and more pronounced tubercles than *M. delawarensis*. Sean Newman.

Fig. 7 Large and well preserved *Menabites (Delawarella) delawarensis* from the "red zone" of the Ozan Formation upstream from the FM 2990 bridge. This is the more common species in the Ozan. Chris Vencevich.

Figs. 1, 2 Two views of a very large specimen of *Pachydiscus paulsoni* from the "red zone" within the Ozan Formation. This genus has the most complex suture pattern of any common ammonite from the NSR study area. Duncan MacKenzie.

Figs. 3, 4 Two views of an unusual specimen of *Pachydiscus paulsoni* within a nodule from the Ozan Formation. The complex suture pattern is clearly demonstrated here. Richard Benefield.

Figs. 5, 6, 7 Three different specimens of *Trachyscaphites spiniger spiniger* from the "red zone" of the Ozan Formation. This form is tightly coiled in its early stages and opens up in more mature forms. Tubercles are pointed. Figs. 5 & 6 are macroconchs and Fig. 7 is a microconch. Fig. 5 by Bill Jolley, Fig. 6 by Larry Shindel and Fig. 7 by Ron Morin.

Figs. 1, 2 two views of bulk sampled microteeth of the triggerfish *Stephanodus* aff. *minimus* from the basal unit of the the Ozan Formation along Moss Creek. Mark McKinzie.

Fig. 3 A complete left prearticular pycnodont dentition from the Ozan Formation. Probably from Anomoeodus barberi.
Dennis Magar.

Figs. 4, 5, 6 Three separate specimens of the frontal palatine teeth and upper jaw section of *Enchodus petrosus*. The "sabre-tooth herring" of the NSR study area.
Mark McKinzie.

Fig. 7 Large teeth from the lower jaw of *Enchodus petrosus*.

Fig. 8 Large, flattened conical teeth of *Protosphyraena sp.* Ed Swiatovy.

Fig. 9 Large, well preserved, spool shaped vertebrae of predatory fish *Xiphactinus audax* from the Ozan Formation. Ed Swiatovy.

Figs. 10, 11 Two views of a complete dentary (lower jaw section) of *Saurodon sp.* from the Ozan formation. The front of the jaw would have extended to the right where this part connected to the pre-dentary. Ed Swiatovy.

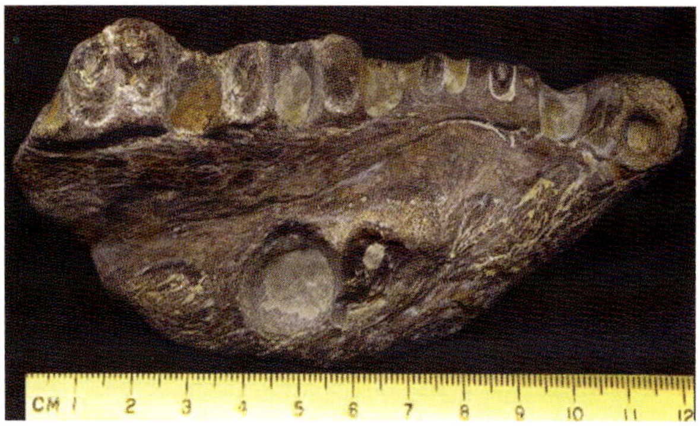

Figs. 1, 2 Lower and upper views of the pterygoid (jaw section) of *Pachyrhizodus caninus*. This predatory fish is characterized by its long and heavy jaws. They are commonly mistaken for marine reptile remains. Collected from the Ozan Formation. Valerie Cheshier.

Figs. 3, 4 Right and left lateral views of a crushed but articulated skull of *Pachyrhizodus minimus* from the Ozan Formation. Note the sharp, short conical teeth along the length of the jawline. Don and Jane Fagerstrom.

Fig. 5 An articulated pectoral fin fragment of *Protosphyraena sp.* from the Ozan Formation. Gerry Hightower.

Fig. 6 Partial skeleton of *Hoplopteryx sp.* from the Ozan Formation. Valerie Cheshier.

Figs. 7, 8 Two views of a laterally compressed partial skeleton of *Hoplopteryx sp.* from the Ozan Formation. This piece possibly represents the remains of a predator's meal that sank to the bottom without being digested. From the Gerry Hightower estate.

Fig. 1 Exterior view of a nearly complete left dentary (lower jaw) of the shell crushing mosasaur *Globidens alabamensis* from the Ozan Formation. Front of the jaw to the left. There are 7 peg-like teeth visible including one tooth cropping out along the side of the jaw. Gerry Hightower.

Fig. 2 Exterior view of another *Globidens alabamensis* from the Ozan Formation. Note the jaw muscle scar attachment sites extending horizontally along the median section of this specimen. The same specimen is figured in Fig. 4.

Fig. 3 Interior view of the same *Globidens alabamensis* left dentary as in Fig. 1. Length of the specimen is 11 inches. There are a total of eight teeth and one socket along this jaw section.

Fig. 4 Interior view of another partial *Globidens alabamensis*. Note how the teeth to the right are more asymmetrical. The front four teeth of *Globidens* are more peg-like than globular. This indicates that this is an anterior portion of the jaw. From the estate of John Moody Sr..

Fig. 5 Occlusal view of the jaw section in Fig. 2 & 4. Note the two tooth sockets to the right. These are the frontal two teeth on this jaw. From the estate of John Moody Sr.

Fig. 6 A tooth series from different individuals of the mosasaur *Tylosaurus* showing the variability in size and shape depending on location within the jaw. Recovered from the Ozan Formation. Gerry Hightower collection.

Fig. 7 Close-up of an singular *Globidens alabamensis* tooth. Note the length of the root. Dennis Mayar.

Fig. 8 Close-up of an individual *Tylosaurus sp.* tooth. Note the fine striations running the length of the blade.Gerry Hightower.

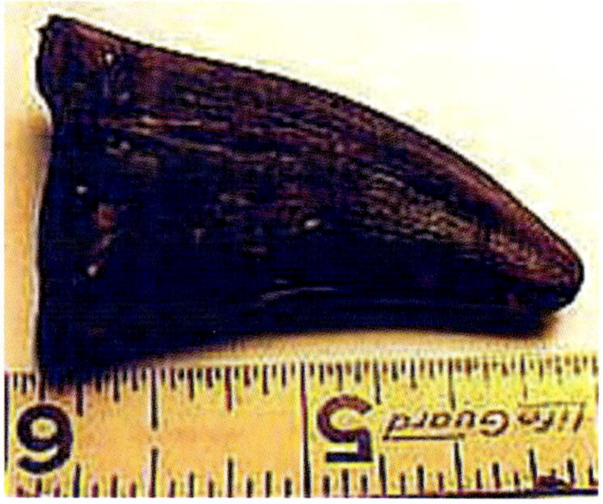

Fig. 9 Closeup of a *Tylosaurus sp.* tooth from the Ozan Formation by Ed Swiatovy.

Partial Skull of the Mosasaur *Clidastes liodontus*

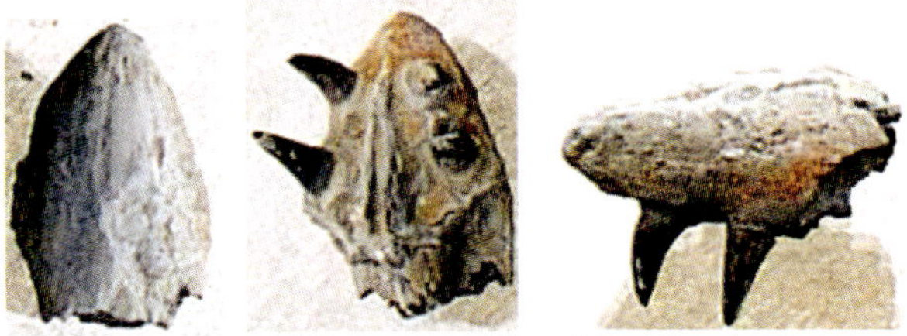

Figs. 1, 2, 3 Dorsal(top), ventral(inside) and right lateral views of the **Premaxillary** (snout = rostrum).

Fig. 4 Lateral view of the right **Maxillary** (upper jaw). Note ten teeth and four empty socket spaces along the jaw.

Figs. 5, 6 Dorsal (top) and ventral (underside) views of the **Frontal** bone -large triangular bone between the eye sockets at the top of the skull.

Figs. 7, 8 Dorsal (top) and ventral (inside) views of the **Parietal** bone. Attaches to the Frontal behind the eyes.

Figs. 9, 10 Labial (outside) and lingual (inside) views of the right **Articular** (jaw hinge) – towards the back of the skull.

Fig. 11 1st **Pygal** (neck)vertbrae behind the skull.

Fig. 12 Round (stone found with it gastrolith)?

Figs. 13, 14 Right lateral and ventral views of the last neck vertebrae. Species identification by Mike Everhart.

Fig. 15 Two more associated vertebrae found with the partial skull. All specimens collected by Ed Swiatovy from the Ozan Formation along the NSR

Figs. 1, 2 Rostrum in right lateral and interior view of a very large *Tylosaurus proriger* collected by Don and Jane Fagerstrom from the Ozan Formation. Twelve inch ruler for scale. Features distinguishing *Tylosaurus* from the more common *Clidastes* includes the larger overall size and the more "bulbous" and flattened tip of the rostrum. In addition, the first set of teeth occur further back in the jaw than those present in *Clidastes* (compare with the *Clidastes* from Plate 13, Fig. 3). You can easily imagine the Tylosaur rostrum being used as a battering ram to stun prey. The word *Tylosaurus* is a combination of Greek for *tylos* "protuberance" + *sauros* "lizard".

Figs. 3-5 Dorsal view, anterior view and close-up interior view of the rostrum. Note the unerrupted juvenile tooth waiting to take its place beneath the exposed tooth socket.

Figs. 6, 7 Two views of the associated neck (cervical) vertebrae from the skull shown above. A yardstick for scale. The back of the skull would have been oriented to the right in both pictures. Note the large size of each vertebrae.

Figs. 8-10 Three views of a individual tail (caudal) vertebrae with the processes still attached. Twelve inch ruler for scale. Numerous bones show the tell-tale evidence of shark scavenging in the form of scratch marks. Two *Squalicorax* teeth were found imbedded in two separate bone specimens from this individual.

Figs. 1, 2 Dorsal and ventral views of a caudal vertebrae of *Halisaurus sp.* from the Ozan Formation. Figs. 1-3 Richard Benefield.

Fig. 3 Left lateral view of a lumbar vertebrae of *Halisaurus sp.*

Fig. 4 Mosasaur vertebrae with the spinal processes still attached. Possibly *Clidastes sp.* From the Ozan Formation.

Figs. 5, 6 Side and top views of a very large plesiosaur vertebrae from the Ozan Formation. Note the overall spindle shape of the vertebrae versus mosasaur vertebrae. Figs. 4-6 Richard Benefield.

Fig. 7 Three associated plesiosaur vertebrae From the Ozan Formation. John Moody Sr.

Fig. 8 Very large plesiosaur vertebrae from the Ozan Formation. Clay Dorris.

Fig. 9 Dorsal view of the right humerus from the pectoral girdle of a *Clidastes sp.* from the Ozan Formation. John Moody Sr.

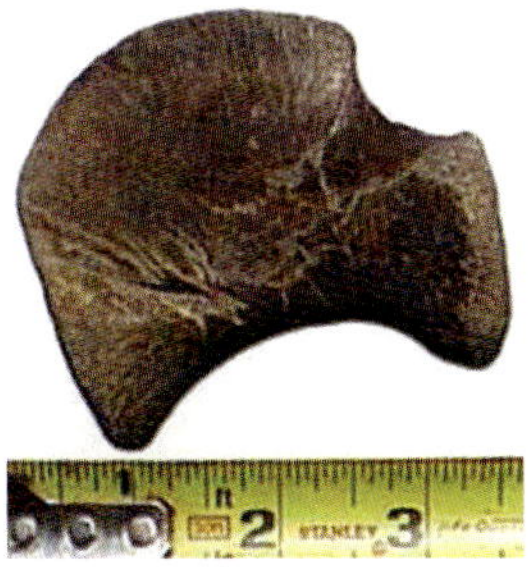

Figs. 10-13 Skeletal elements from the pectoral (shoulder) girdle of *Clidastes sp.* from the Ozan Formation. Fig. 10 is the ventral aspect of the left radius. Fig. 11 is the dorsal aspect of the right scapula. Figs. 12 and 13 are the ventral and dorsal views of the right radius. John Moody Sr.

Figs. 1, 2 Two views of a magnificent *Clidastes liodontus* skull collected by Lewis Smith in October, 1991 from the Ozan Formation. Fig.1 is how it was appeared *in situ* (and underwater!). Fig.2 is the reconstructed skull on display at the Greenville Christian School in Greenville, Texas.

Fig. 3 Large mosasaur vertebrae with a deep gouge from a predator and/or scavanger. Ed Swiatovy.

Fig. 4 Pectoral girdle element of a protostegid predator (marine) turtle from the Pecan Gap Chalk downstream from Hwy 24. Mac Glaess.

Figs. 5, 6 Assorted scutes from the hypoplastron and carapace of a Cretaceous marine turtle. Identification is not possible based on these individual scutes only. From the Ozan Formation. Mark McKinzie.

Figs. 7, 8 Two views of unidentified dinosaur bones. These bones are too round and large to have come from a marine reptile and represent the remains of a dinosaur carcass that floated out to sea. Mark Randall.

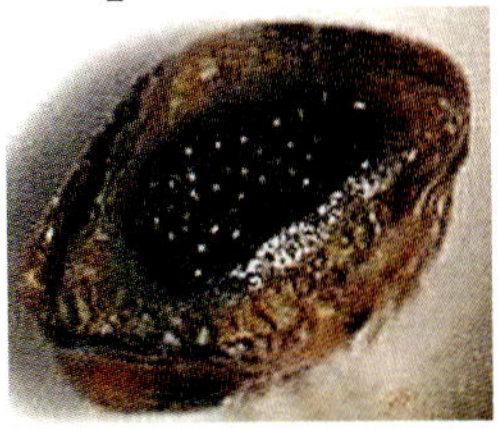
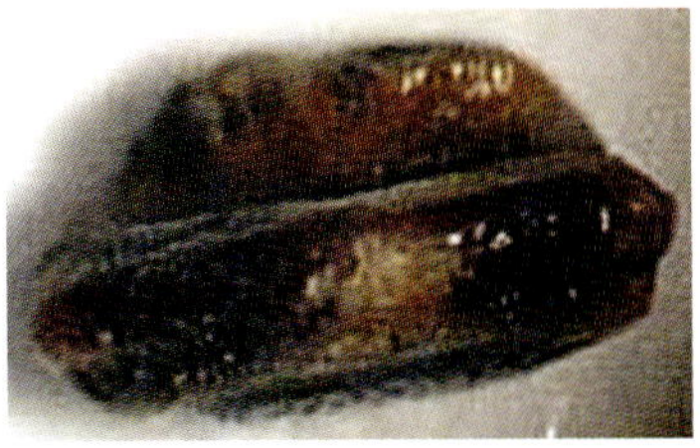

Figs. 1-4 Assorted microteeth from the basal Ozan Formation beneath the Hwy 34 bridge over Moss Creek. Figs. 1, 2 Occlusal and basal views of *Myliobatis sp.* Fig. 3 Side view of *Rhinobatis ladoniaensis.* Fig. 4 Occlusal view of *Heterodontus sp.* All specimens Ed Swiatovy. Magnification x10.

Figs. 5-7 Occlusal (top), side, and basal (root) views of a large *Brachyrhizodus wichitaensis* from the gravel bars underneath the Hwy 24 bridge over the NSR. Mark McKinzie.

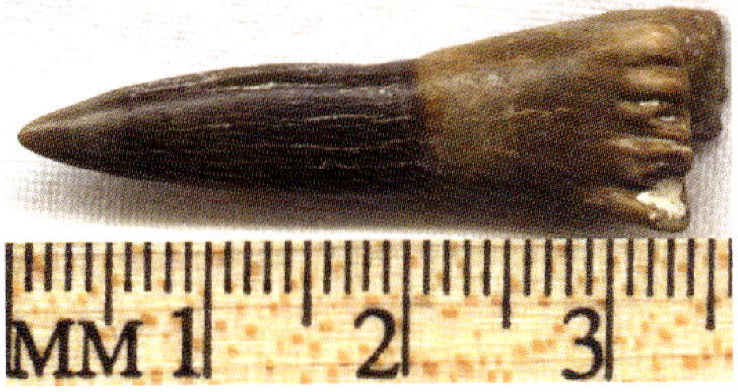

Fig. 8 Well preserved specimen of the rostral tooth of *Ischyrhiza mira.* From the Ben Franklin RR tracks locality. Mark McKinzie.

Fig. 9 Rarely preserved cartilage section with denticles. Mark McKinzie.

Fig. 10 Beautiful example of a shark corprolite. Note the spiral shape. Valerie Cheshier.

Figs. 11-13 Three views of another shark with the preserved dermal layer ("shagreen") draped over the three associated vertebrae. This is a very unusual and rare specimen. Don and Jane Fagerstrom.

Figs. 14, 15 Two views of "platter" size shark vertebrae from the Ozan Formation. Don and Jane Fagerstrom.

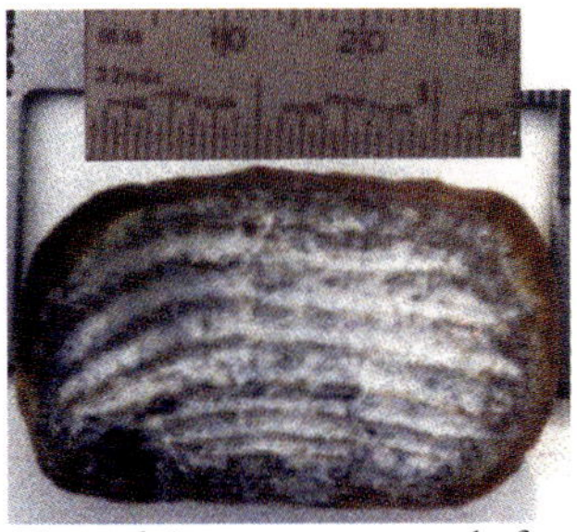

Fig. 16 A pavement tooth from the rare *Ptychodus connlyei*. Roger Farish.

Fig. 1 Assorted dermal scutes (bony armor) of *Alligator mississippiensis*. Mark Mckinzie.

Figs. 2, 3 Top and oblique views of an armored tail ring from the armadillo *Dasypus bellus*. Marquita Westerman.

Fig. 4 Nearly complete incisor from the giant beaver *Casteroides ohioensis*. Found by Mac Glaess.

Fig. 5 Canine tooth from the black bear *Ursus americanus*. Jim Chaney.

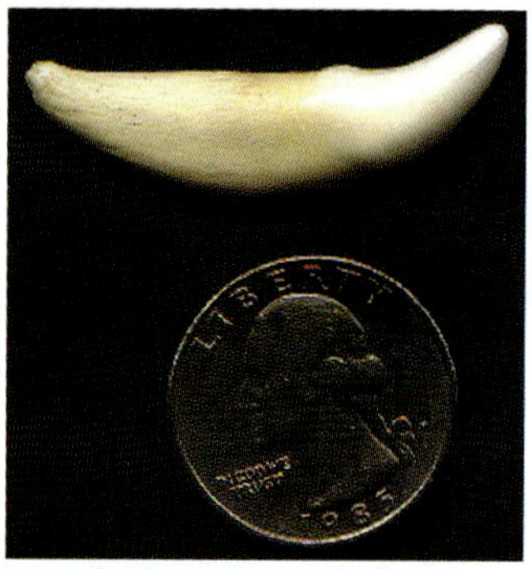

Fig. 6 Canine tooth from the coyote *Canis latrans*. Valerie Cheshier.

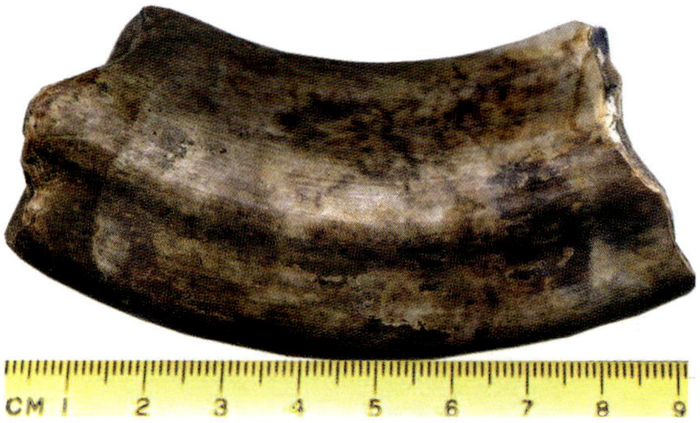

Fig. 7 Partial (broken at both ends) incisor to the giant ground sloth *Megalonyx jeffersoni*. Mac Glaess.

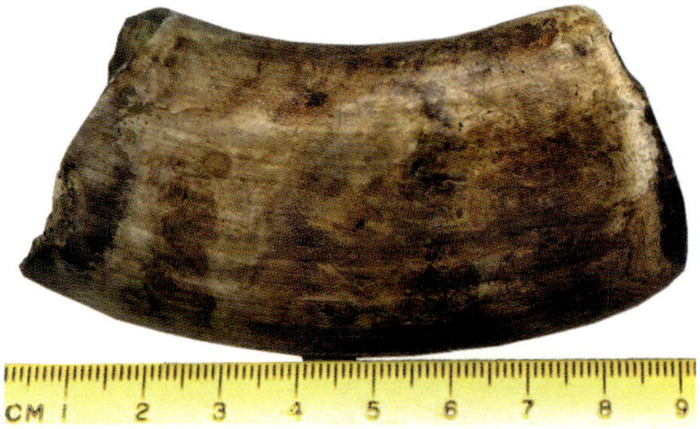

Fig. 8 Another partial incisor to the giant ground sloth *Megalonyx jeffersoni*. Rick Kaiser.

Figs. 9, 10 Two views of the ankle bone (astragalus) from the horse *Equus sp.* Mark Randell.

Fig. 11 Oblique view of the astragalus from the ground sloth *Megalonyx jeffersoni* Dennis Magar.

Fig. 12 The inner pulpy core from the thumb claw of *Megalonyx jeffersoni*. Mark Randell.

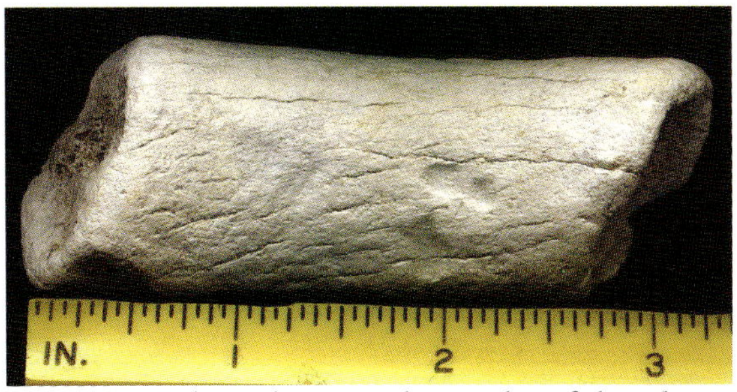

Fig. 1 Partial deer jaw with four molars of the white-tail deer *Odocoilus virginianus*. Mac Glaess.

Fig. 2 Partial weatherworn deer antler of the white-tail deer *Odocoilus virginianus*. Valerie Cheshier.

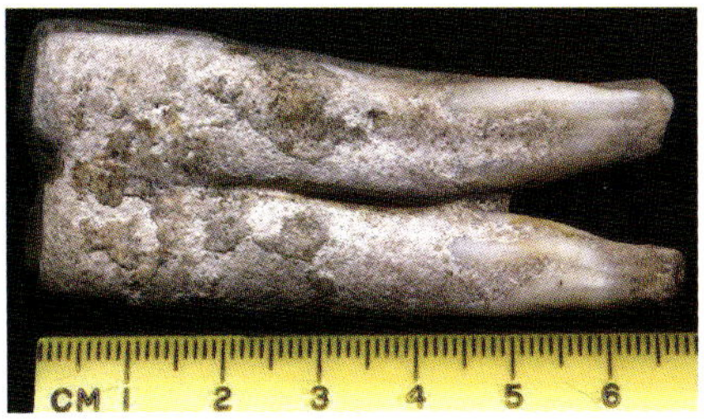

Fig. 3 Molar crown from the tapir *Tapirus* Larry Shindel.

Figs. 4, 5 Two different molars from the modern horse *Equus sp.* Note the distinctive elongated bilobate root. Chris Vencevich and Mark McKinzie

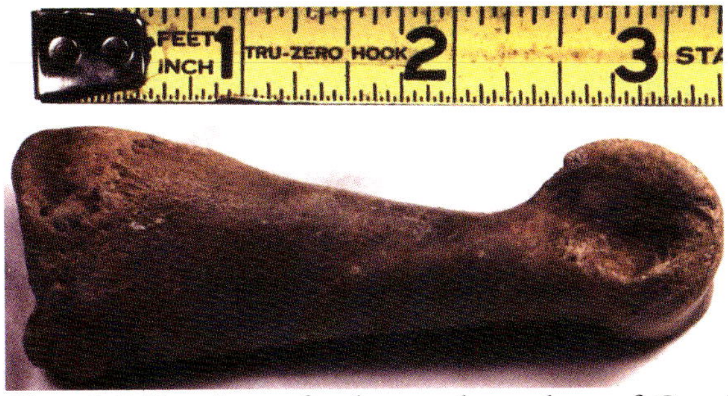

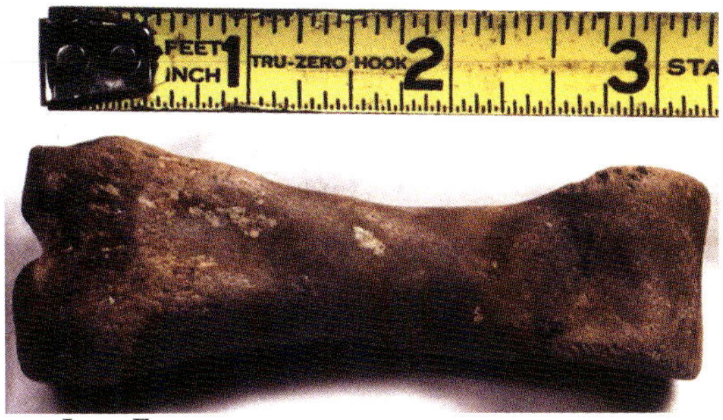

Figs. 6, 7 Two views of a elongated camel toe of *Camelops sp.* Jane Fagerstrom.

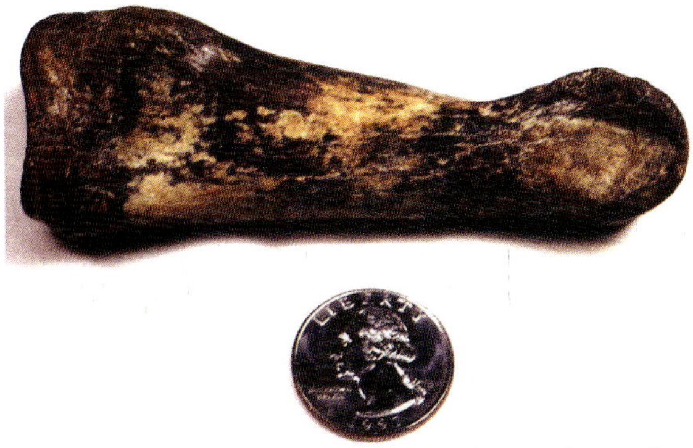

Fig. 8 Another view of a different elongated camel toe *Camelops sp.* Dennis Magar.

Fig. 9 An unidentified hoof core from a bone from Pleistocene quadraped. Possibly a deer. Bob Wilk.

Figs. 10, 11 Two views of the small body scutes from the armored pampathere *Holmesina septentrionalis*. From the John Moody Sr. collection.

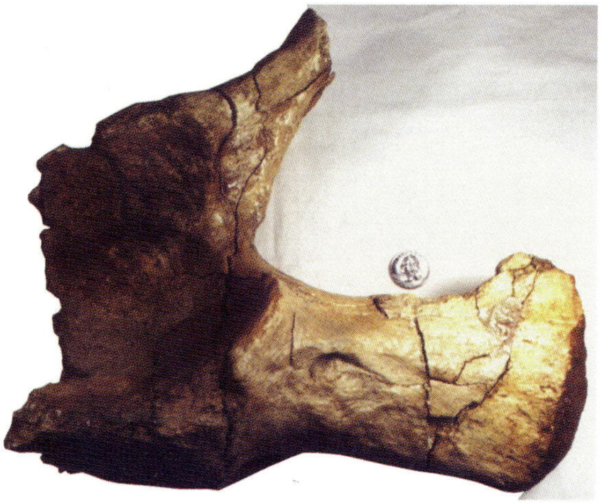

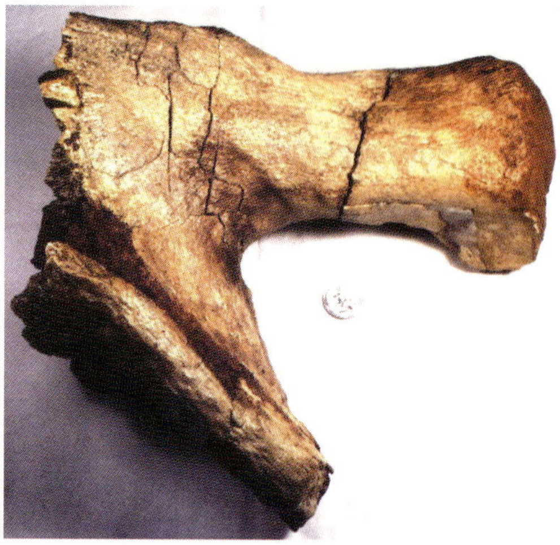

Figs. 1, 2 Two views of a partial pelvis from a juvenile mammoth, probably *Elephas cf. columbia*. Note the quarter for scale. Mac Glaess

Figs. 3, 4 Two views of an individual rib from the Columbian mammoth *Elephas cf. columbia*. Note the quarter for scale. Mac Glaess.

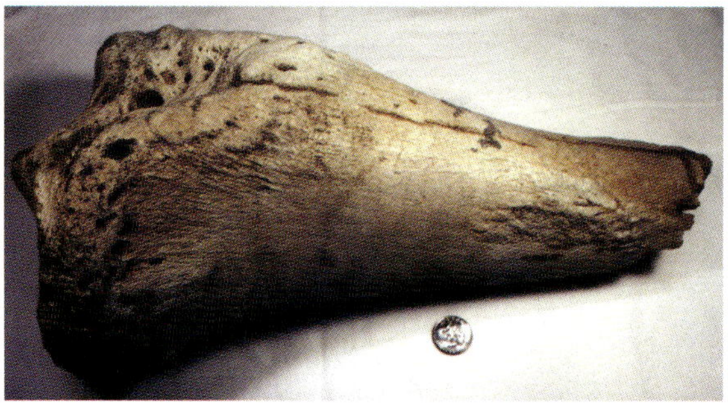

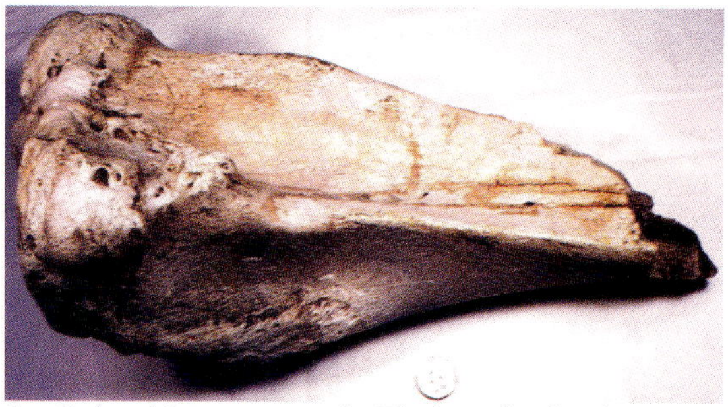

Figs. 5, 6 Two views of the spinal process off a vertebrae from the Columbian mammoth *Elephas cf. columbia*. Note the quarter for scale. Mac Glaess.

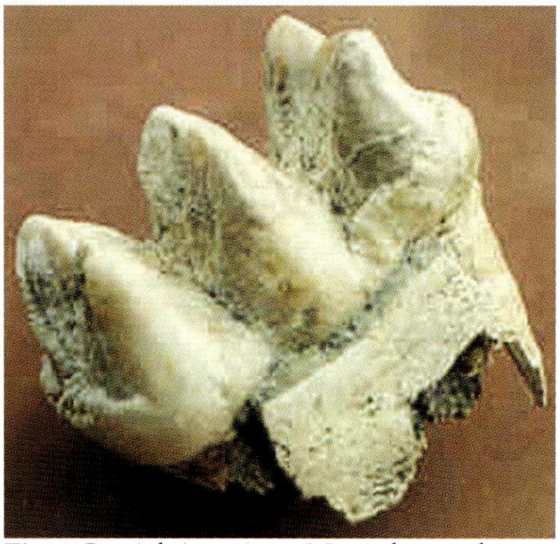

Fig. 7 Partial American Mastodon molar *Mammut americanus*. Lewis Smith.

Fig. 8 Nearly complete Woodland musk ox skull, *Symbos cavifrons*. It can be distinguished from the extant muskox *Ovibos* by the rough-pitted surface between the horn cores. Lewis Smith and Harold Finley.